KU-754-649

To Richard, my family and my friends.
For always believing in me.

SCIENCE AND THE CITY

THE MECHANICS BEHIND THE METROPOLIS

Laurie Winkless

Bloomsbury Sigma
An imprint of Bloomsbury Publishing Plc

50 Bedford Square
London
WC1B 3DP
UK

1385 Broadway
New York
NY 10018
USA

www.bloomsbury.com

BLOOMSBURY and the Diana logo are trademarks of Bloomsbury Publishing Plc

First published 2016. Paperback edition 2017.

Copyright © Laurie Winkless, 2016

Laurie Winkless has asserted her right under the Copyright, Designs and
Patents Act, 1988, to be identified as Author of this work.

All rights reserved. No part of this publication may be reproduced or transmitted
in any form or by any means, electronic or mechanical, including photocopying,
recording, or any information storage or retrieval system, without prior permission
in writing from the publishers.

No responsibility for loss caused to any individual or organisation acting on
or refraining from action as a result of the material in this publication
can be accepted by Bloomsbury or the author.

Every effort has been made to trace or contact all copyright holders.
The publishers would be pleased to rectify any errors or omissions
brought to their attention at the earliest opportunity.

British Library Cataloguing-in-Publication Data
A catalogue record for this book is available from the British Library.

Library of Congress Cataloguing-in-Publication data has been applied for.

ISBN (paperback) 978-1-4729-1323-4
ISBN (ebook) 978-1-4729-1322-7

2 4 6 8 10 9 7 5 3 1

Chapter illustrations by Neil Stevens
Diagrams by Marc Dando (unless credited otherwise)

Typeset by Deanta Global Publishing Services,
Chennai, India

MIX
Paper from
responsible sources
FSC
www.fsc.org   FSC® C020471

To find out more about our authors and books visit www.bloomsbury.com.
Here you will find extracts, author interviews, details of forthcoming
events and the option to sign up for our newsletters.

Contents

Begin

I am well aware of how annoying I am.

Ever since I learned how to talk, it's been impossible to stop me from asking questions. Those kind, generous souls amongst you might say that I'm just endearingly curious, and I thank you for that. But I'm OK with admitting that I just get a kick out of finding answers. I think it was this that led me to study physics, and motivated me to go into research. In a weird way, it's why I'm now sitting here, writing this introduction. A while ago, I realised that there were countless interesting questions to be asked in the very place where I live; where many of us live.

Cities

In 2014, the UN announced that, for the first time in human history, more than half of the world's population now live in urban areas. We are officially no longer farmers; we are urbanites. Cities have never been bigger, busier or more important. Luckily, for the past 11 years, I've lived in what I think is the greatest city in the world: London (feel free to argue with me – I promise you, my love is of an unshakeable kind). For the first little while, I was like a rabbit in headlights. But gradually, once the initial shock of the scale of the city wore off, I started to notice things, such as strange pipes hidden in the dark walls of tube tunnels. I found myself wondering how it all tied together, and what kind of engineering, technology and science lay behind it. In short, I wanted to figure out how cities work, and to help others to do the same.

As it turns out, all I needed to do was to spend too much time on Twitter, attract the attention of a publisher and then convince them to let me write for them… Easy! Two years (and more high-vis than I ever could have imagined) later, here it is, *Science and the City* (*SATC*) – my very first book. It's basically my scientific love-letter to the great cities of the world.

In *SATC*, I will take you on a big, sciencey exploration of your city. Through a series of jam-packed chapters, we'll peek behind doors and under the footpaths in cities across the globe, to discover the secrets therein. We'll talk about everything from transport networks to the systems that get water and electricity to you (separately!). With me as your direct line to hundreds of experts, you'll find answers to your questions about the urban jungle. I'll explain how the technology all around us works, and help you to build your own vision of the cities of tomorrow. Together, we'll trawl through the latest research papers to find the coolest technologies you've never heard of, which are set to change the way we build, travel and work in the future. It's going to be fun.

This is *not* a book written by a (lapsed) scientist solely for other scientists. This is for everyone who has ever wondered how traffic lights work, or why birds can perch safely on power cables. *SATC* is for those people who are curious about the world they live in.

Throughout this process, I've met some remarkable scientists and engineers, people who have genuinely changed the way I see the world. But the real stars are the cities themselves – so to them I say, thank you.

CHAPTER ONE

Up

Close your eyes. Imagine that you're standing in one of the world's financial hubs – say London, Hong Kong or New York. Breathe in your surroundings and describe the scene. Perhaps you've found yourself next to one of the city's landmarks; maybe you're enjoying a coffee on a side street, or even battling your way through traffic in a cab. No matter where you are in the urban jungle, one thing's for sure: somewhere in your mind's eye are numerous shiny, towering buildings of steel and glass – **skyscrapers**. Few words better sum up a vision of a vibrant, growing, future-facing city, so they're a perfect place to begin our exploration.

Skyscrapers aren't a truly modern invention. High-rise buildings have been found in the heart of cities for 130 years, and the motivation behind constructing them is older still. In areas like Hong Kong Central or Lower Manhattan, there's a limited amount of land but plenty of people who want it, so the only way is up.

Today

How much stuff can we fit onto a piece of land? Well, in 2013, the average American home had a footprint of 232m² (2,500sq ft), and that housed three people. In the heart of New York's financial district, it's a little different. Each tower of the now-destroyed World Trade Center (WTC) had a footprint of just over 4,000m² (43,000sq ft) – or about the size of 17 average homes. But instead of housing 51 people, every day each of the WTC towers accommodated 25,000 of them! This isn't really a fair comparison, of course. Most office blocks don't attempt to offer living space; they only offer working space. Without the need to provide a bed for every worker, the area required to accommodate them shrinks dramatically. But the big picture remains – because we can create tall buildings, a relatively small area of land can be comfortably used by thousands of people.

For me, the skylines that these buildings create represent some of our very best engineering. Skyscrapers must do a bit of everything: they support their own weight and that of the stuff they house, they resist wind (and, occasionally, earthquakes) and protect occupants from fires or floods. They must also be eye-catching, and provide a spacious, pleasant environment, with all floors easily accessible. Oh, and they have to be fully connected to the power grid, and offer a reliable water supply, waste removal infrastructure and communication systems. Doing all this while still realising the original vision of the architect is not easy; it requires a huge amount of engineering skill and considerable planning.

Be honest – have you ever wondered how we can even build such huge towers? I guess the first question to ask is when does a tall building become a skyscraper? The Council on Tall Buildings and Urban Habitat (CTBUH) – keeper of the world's skyscraper info – does have some definitions in place. If a building is taller than 300m (around 1,000ft), it is called supertall; if it reaches above 600m, it is called megatall. But we can broadly think of a skyscraper as being a building that is *much taller than it is wide*. In order to figure out how we've arrived at the modern vision of high-rise cities, I'd like us to build a skyscraper from its raw ingredients. We'll talk about the materials involved, as well as the equipment and building techniques that will ensure our skyscraper stays upright. There are plenty of challenges to overcome along the way. To get started, let's take a step back in time to the Industrial Revolution.

Steel

Until the mid-1800s, large buildings were made of four standard materials: wood, stone, brick and iron. Wood was our first true construction material, and it's never really gone away. For bigger structures, though, stone and brick dominated for millennia, but they required a lot of preparation – stone had to be machined into blocks, and bricks were hand-formed and baked in a furnace. Both were incredibly heavy, which limited how tall buildings could be before they began to collapse under their own weight.

Wrought iron and cast iron became relatively easy to produce in the 1800s, and they offered the visionary builders of the time a new way to look at construction. The Eiffel Tower in Paris was built using wrought iron, and weighed in at 7.3 million kilograms. In order to truly scrape the sky with our structures, we would need to find lightweight, strong materials that could be made cheaply.

The first step was to move from iron to steel. Metallic iron is one of the most common ingredients of the Earth's crust, which made it easy to find even in prehistoric times.[*] But in its raw form, iron is rather soft. To make it suitable for construction, we must add a little something extra to make it an alloy.

- **Pig iron** contains lots of carbon (3.5–4.5 per cent) along with other impurities. It is soft and very brittle, not particularly useful for a construction material, but we'll come back to it.
- **Cast iron** contains slightly less carbon (2–3.5 per cent) as well as up to 2 per cent silicon, which makes it hard but brittle, meaning it can break under certain types of stress.
- **Wrought iron** contains a tiny amount of carbon (0.02–0.08 per cent) and the result is a hard but malleable metal that can be hammered, rolled or pressed into sheets without breaking.

---

[*] Just like us, iron is made of stardust. Stars get their brightness by a process called nuclear fusion – the slamming together of atoms to make heavier atoms. Everything you see on the Periodic Table from hydrogen to iron was created inside the heart of a star. Iron *can* fuse to form heavier elements, but that process uses more energy than it produces.

Between these last two sits the alloy we want: Steel, or rather, steels, contain anywhere between 0.5 and 2 per cent carbon, making them harder than wrought iron, yet malleable and flexible, unlike cast iron. So these are potentially perfect for our needs. However, in the early days of the Industrial Revolution, steel was extremely expensive to produce – upwards of $80 (£57) per tonne – and its production was hugely labour intensive.

In 1855, everything changed. It was then that a British inventor called Henry Bessemer patented a new process for mass-producing steel. It involved pumping compressed air into a large pear-shaped vessel filled with molten pig iron. The idea was that the oxygen in this blast of air would remove some of the carbon in the pig iron (by forming carbon dioxide, CO_2), leaving a molten steel. Good in theory, utterly unreliable in practice. The other impurities in the pig iron meant that Bessemer could never really be sure of the composition of the final steel, and his growing customer list became increasingly angry (some even sued him). It took the input of Robert Mushet, a British metallurgist, to save Bessemer. His suggestion to remove all of the carbon first and then add small percentages back in later drastically improved the process.

Bessemer's system rapidly began to change the world of steel manufacturing, and by 1875, costs had dropped to $32 (£23) per tonne. As always, in the supply-and-demand equation, the availability of cheap, high-quality steel made it immensely popular, leading to another huge drop in the price per tonne. We'd entered the modern steel age, and we haven't left it yet.

Concrete

So now, back to our skyscraper. With steel, we have a material that is cheap and reliable, and can be used to build a tall frame. But what about the rest of the building? You may not think of it as such, but concrete is a high-tech material, and it has been around a lot longer than cheap steel. In its simplest form, concrete consists of small particles of a hard material (known as aggregate) bonded together by a cement and water, and these days we can even reinforce it using steel bars. But it came from humble beginnings. The ancient Babylonians built structures with a mixture of clay and pebbles, and parts of the Great Wall of China used cement-like material to bind blocks together. 'Modern' concrete only truly emerged in the mid-1800s, with the invention of Portland cement.

Its first engineering application was to fill a breach in a tunnel under the River Thames in London. If you want to make your own cement, it's not enough to grind up just any old rocks – you need rocks that contain calcium carbonate (e.g. limestone) and silicates (silicon + oxygen). For Portland cement you'll also need a pinch of iron ore, and a dash of aluminium oxide. These ingredients are then ground up, heated to temperatures above 1,450°C (around 2,640°F), and then ground again to a very fine powder. To make concrete, simply mix this with water and aggregate, *et voilà*!

These days, concrete is everywhere, and that's partly because it has some pretty cool properties. First, by playing around with the choice of cement or aggregate, you can define how strong the final material is, how durable it is or even just how it looks. Because concrete can be poured into moulds, you can make it into complex shapes. It's naturally fire-resistant and is less susceptible to rot, corrosion or decay than other building materials. And, it gets stronger over time.

There are less cool things about concrete too (which we'll talk about in later chapters), but let's just unpick that last point: concrete gets stronger because of the chemistry that occurs between Portland cement and water. Contrary to popular belief, concrete doesn't 'dry out'; it ties water tightly into its structure by a process called curing. When you add water to your fine powder of calcium silicate (which is calcium, silicon and oxygen bonded together), you produce a series of very sticky compounds that cling on to the aggregate, forming concrete. While hydration is clearly important, you still have to be careful – too little water and the reaction doesn't complete; too much and unused water can sit inside the material, weakening it. So rain on a day of concrete-laying can be a problem. Generally on building sites, newly poured concrete is either sprayed with a water mist, or covered in moisture-retaining fabric (such as burlap) while it cures. Once the initial curing is complete, concrete can continue to harden using moisture from the air, until it reaches the point when it is as strong as it's going to get.

'Strong' is one of those words that is used every day, but for scientists it has a specific meaning, or a number of specific meanings. Here are just three:

1. **Compressive strength** is a measure of how much you can squeeze a material before it fails.
2. **Tensile strength** tells us how hard you can pull on a material without it breaking.
3. **Shear strength** represents how hard you must try to cut a material before it yields.

Concrete is strong under compression – most buildings use it in their foundations because it is good at withstanding the 'squeezing' force between the soil below the foundations, and the weight of the building's walls. But it is very weak under tension, when any force acts to stretch it. This is why concrete beams are never used above doorframes. The weight of the bricks above the beam isn't counteracted by anything below it, which causes the concrete to stretch, and eventually crack. When it comes to constructing skyscrapers, cracking concrete could be fatal. In reality, most of the concrete you see on building sites is reinforced concrete that has a steel mesh or a grid of steel rods (called 'rebar' by those in the know) running through it. This addition improves the tensile strength of concrete, soaking up the stress and minimising the risk of cracking. Despite their obvious physical differences, concrete and steel expand and contract at almost exactly the same rate when temperatures vary. This fact is the single reason that concrete can be used in buildings across the world – it makes managing constantly changing temperatures almost easy.

Because of all these considerations, concrete has become truly ubiquitous in the construction industry, and is a favourite of skyscraper-builders the world over. Its mechanical properties mean that it can be used in most load-bearing structures (such as walls and pillars), and when used in flooring, it can be poured directly into place. So, we should probably have some for our skyscraper.

Glass

In terms of the skeleton of our building, I think we're almost there. But what about the skin? Time to talk about glass, because we'll need tonnes of it. Archaeological records suggest that humans first made glass in around 3500 BC, and the main ingredient has stayed much the same since. Glass comprises mostly silica (SiO_2) – in ancient times, it made up about 90 per cent, but these days, SiO_2 is more like 75 per cent of it. The addition of other ingredients helps to tailor the physical properties of the glass; for instance, adding sodium carbonate lowers its melting temperature to a balmy 1,200°C (2,190°F). The addition of lead oxide can increase its reflectivity, and boron oxides can strengthen it to withstand high temperatures.[*]

But how do we produce huge, flat sheets of the stuff? Invented by Alastair Pilkington in the 1950s, the float glass method is now used to produce almost all of the world's flat glass, and without it, our skyscrapers would be very different indeed. Molten glass is poured onto a bath of molten tin, and because glass is less dense, it floats on top. In addition, the tin is not as hot as the glass ('just' 232°C, 450°F), so as the glass spreads it cools, leaving an incredibly smooth surface on both sides. This process is continuous, which makes it possible to produce mind-bogglingly huge sheets of glass. For global glass brand Pilkington United Kingdom (yes, named after Alastair), the maximum size of glass panes they produce is 6 x 3.21m

---

[*] Boron oxides are the secret ingredients in Pyrex glass cookware.

(19.6 x 10.5ft) – large enough to cover an average living room.*
After production, glass can be heat-treated and/or coated to
further refine its properties.

'Traditionally, it is toughened glass that has been used for
skyscrapers,' said Phil Brown from Pilkington. It is four or
five times stronger than ordinary glass, thanks to a tightly
controlled thermal treatment that forces the glass molecules
to compress together. Because of this strength, it can span
large distances with very little support. But as buildings have
become taller, architects and engineers have demanded more
than strength. 'Now we're moving towards a hybrid glazing –
two sheets of toughened glass with a layer of polyvinyl
butyral (PVB) sandwiched in between.' In the unlikely event
of this toughened-laminated glass being damaged, the sticky
PVB layer will ensure that the pane remains intact. Brown
also told me that almost all of the glass used in tall buildings
is high-performance solar control glass, coated to keep the
heat outside in summer and inside in winter. The coatings
needed to achieve these properties are added to the inner
surfaces of a double-glazed window, protecting them from
environmental damage, and ensuring they last as long as the
window does.

One thing that's worth mentioning is that the huge glass
facades in many tall structures don't actually support any other
part of the building. They just kind of *hang* from the outside of
it, like glass curtains. Following the noble engineering tradition
of giving things incredibly obvious names, these facades are
called **curtain walls**. Now, that's not to say that glass windows
and their frames don't add anything to the building – they
certainly add mass – but they don't have a direct role in keeping
the building upright. And they do look rather gorgeous, so we
definitely need some for our construction project.

---

* This maximum size is mostly set by logistics – shipping anything
larger would be difficult, and most of the furnaces used to heat-treat
glass measure around 4 x 2m (13 x 6.5ft).

Construct

Now that we've collected enough materials to build our skyscraper, we need to agree on its height. Given the one-way nature of our conversation, I'll have to assume you're happy with a 90-storey structure (approximately 380m, 1,247ft). Throughout the building process we'll need to consider the weight of the structure, called the dead load, and the weight of anything it will contain (humans, office equipment, elevators...), known as the live load. The foundations of a skyscraper need to support both the dead load and the live load, so let's start there.

On a hard surface, foundations are typically formed by a network of steel beams called piles, which spread like the roots of a tree to distribute the building's weight. For less ideal ground conditions, there are alternatives: the 88-storey, 452m (1,480ft) tall Petronas Towers in Kuala Lumpur are actually built on two concrete rafts supported by piles, some of which extend 110m (360ft) into the ground, in order to reach bedrock. To keep it simple, let's build our skyscraper on nice, solid clay, and use high-performance concrete and steel piles to form our foundations.

So now, the skeleton. The classic image of a skyscraper is a steel frame with vertical and horizontal supports throughout – basically, a very tall steel box split up into smaller boxes. This structure is a good option because it is simple and relatively straightforward to build, but there are still things to consider. The major one is that as our skyscraper gets taller, we'll need more material. To ensure the frame remains supported as we build up, the steel beams will need to get closer together. More beams means a heavier building, and then deeper and wider foundations to support it... and so on. To be practical, a steel frame can only be used for buildings up to 40 storeys tall. Anything beyond that and you'd end up with more steel than floor space. For our 90-storey beast, we need to look at something else.

Hong Kong may be the 'tallest' city in the world, but arguably, Dubai is now the city most synonymous with skyscrapers. At the time of writing, its superstar building is the Burj Khalifa, which stretches 828m (2,716ft) into the desert sky – twice as tall as an Olympic running track is long. I wanted to find out more about the techniques that can be used to build these super-structures, so I spoke to William (Bill) Baker, the chief engineer of the Burj Khalifa, and a partner in the Chicago architectural firm Skidmore, Owings and Merrill. The Burj is an engineering marvel, as well as a beauty, so I found myself a little star-struck to be talking to this modest engineer. But when I realised he spoke almost as quickly as I do, I knew we'd get on well.

Topping out at 163 storeys, the Burj leaves our building in the shade. Skeleton-wise, it's fascinating too – concrete is used from the ground to the 156th level, with steel from there up. This composite structure is part of the reality of supertall buildings; no single material can do everything. Baker told me, 'We normally try to keep a skyscraper as stiff as possible. This limits our materials choices and certainly shaped our design for the Burj.' Key to achieving the required stiffness was a unique structure Baker designed for the building – a buttressed core. This 11m (36ft)-wide hexagon of high-performance reinforced concrete is the Burj Khalifa's supportive spine, allowing it to get taller without a cumbersome steel frame. It is also the only part of the building that stays a constant size and shape from the ground floor to the very top. 'To go higher than ever before, we knew we needed a stiff axle through the centre of the building, and that a closed shape would be much stiffer than an open one.' But being so narrow, the closed hexagonal core was still too slender to reach great heights without help – just think about how stable a single piece of rigatoni pasta would be if stretched to a few metres! So, Baker looked for inspiration in the great gothic cathedrals of old.

When viewed from above, the Burj Khalifa looks a bit like a tripod, with its three 'wings' joined by the hexagonal core. Along the length of each of the wings are so-called corridor walls – these support the core in the same way as buttresses held (and still hold) up the thick stone walls of cathedrals and castles the world over. Coming off these buttresses are the web walls that give the building additional strength, and avoid any torsion (twisting). According to Baker, '... the further you are from the centre, the more effective the walls are in resisting motion. The end walls on the lowest floor of each of the building's three "noses" are thicker than all the rest.'

This combination of a stiff central core and walls of varying thickness provided a new way to build skyscrapers – one that heralded the dramatic height increase seen in our youngest cities. The Y-shaped structure had another key role to play, as we'll learn shortly, and the concrete spine now houses the building's many elevators. In addition, the core made the construction process easier: because its width remains constant, the floor segments immediately surrounding it could be repeated over almost its entire height, and it's much easier to mass-produce a concrete slab than to make a bespoke one each time.

Baker believes that the buttressed core system can take skyscrapers even higher. 'We could have gone beyond 1km

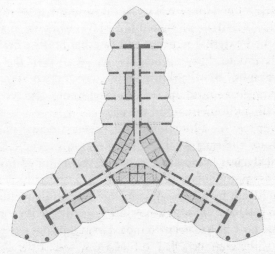

Figure 1.1 The hexagonal core and the tripod-like structure of the building can be seen in this example floor plan of the Burj Khalifa (Credit: Skidmore, Owings & Merrill LLP).

(3,280ft) with the Burj – the limits were really financial, but also to do with the change in air pressure on one's inner ear in going to such great heights.' His response would suggest that Baker dreams very big dreams indeed – for humans, atmospheric pressure only really becomes an issue 2.4km (8,000ft) above sea level, three times higher than the Burj Khalifa. If we were ever to attempt to build a structure that tall (which, frankly, looks rather unlikely), the top floors would need to be pressurised, in the same way that aeroplane cabins are.

Back to our very own homemade skyscraper. We now know that we have options in terms of the skeleton and walls of the building, so let's go for a combination of reinforced concrete and steel for the frame. The Burj Khalifa's innovative structure may be beyond our own construction skills, but if we follow the example of Baker's team, we can use identical units that repeat throughout the structure, to make the build simpler and more efficient. But how do we go about assembling a building that towers hundreds of metres above the ground? The construction of the steel-framed Empire State Building in the 1930s involved a complex system of incredibly brave workers, manually controlled external winches and internal hoists to raise the steel beams to dizzying heights (we've all seen the iconic photographs). But thankfully, these days, cranes are used to lift completed units, and we can even pump concrete more than half a kilometre up in the air.

Baker told me a little more about their concrete-pumping system for the Burj Khalifa: '[One] reason we could build it so quickly was because the concrete was pumped from the ground up to 600m (2,000ft), using a specially-designed system.' Built by a German company called Putzmeister, the system was basically a high-pressure hose filled with concrete. Because no-one had ever attempted to pump concrete to such heights, the engineers had to find a way to test the concept, and they had a very clever solution. By laying horizontal pipes on the construction site and studying the flow of concrete through them, researchers could estimate the pressure needed to pump concrete upwards. In the end, they used pressures a hundred times larger than that inside a typical family car tyre to safely pump concrete into the sky.* I have to admit that every time I think of this, I picture a cola-mint volcano, but I have been assured that it doesn't look anything like that. Shame, really. For the fully-formed pieces, such as the rebar cage, or the steel bars used to construct the remaining storeys, the only option is to lift them. If you've ever been anywhere near a construction site, you'll know what a crane looks like, and you may even have seen some monstrously tall ones. But cranes almost a kilometre (over half a mile) tall do not exist, so what can we do?

We can use a crane that builds itself. First, the crane needs to be supported so that it can 'grow' without the risk of it toppling, so it's usually directly attached to the partly constructed building using strong steel collars. Once it is safely fixed, the crane's special 'climbing section' comes into its own: this is a metal sheath that surrounds the crane's tower and it can be moved up and down easily. Each time we want the crane to get taller, new tower segments, which I think of as vertebrae, can be added to the steel spine. The sheath acts as a temporary support for the crane while the vertebrae are being added and secured, before being raised again. As the building grows, so too does the crane, and further steel collars are added to tie them together. The crane constructs the building, and the building supports the crane – it's almost poetic.

In the middle of 2015, a Chinese construction company announced that they had built a 57-storey skyscraper in the Hunan province in just 19 working days. One contributor to this high-speed build was the use of identical units throughout the entire building. Think about how much quicker it would be to build a Lego wall with three-brick units rather than singles – their speedy modular method added three floors a

---

* The average pressure in the tyres of a Volkswagen Polo is 2.2Bar.

day to the structure. But while it's a very efficient way to build taller, it could lead to cities filled with identical, bland high-rises… and we don't want that.

Elevate

OK, quick recap. We know that we can build foundations, a steel frame, concrete floors and walls, and glass facades. We can pump concrete to the upper floors, and can lift steel rebar and frames using a self-building crane. So we are all set, right? Nearly. We just have the small matter of moving the eventual tenants around, and if you've ever had to walk up 10 flights of stairs, you might already know what's coming next – the elevator (or lift). In a skyscraper, elevators are more than just a utility. Ahead of advances in materials or equipment, elevators are what make skyscrapers practical, or even feasible. And it's thanks to them that our cities have the skylines they do.

Elevators form part of the structure and design of the building – the taller the building, the more of them you need, and so the bigger the footprint of the elevator shaft. And, as we've seen in the Burj, the elevator shaft is often a vital structural component of a building. They are so important that they must be designed into a building from the earliest architect's sketch – retrofitting a lift is a lot harder. The very first (safe) passenger lift was installed in a New York department store in 1857, and it was pretty simple; effectively a compartment attached to a rope and a lifting mechanism. It also had the very first governor device, which used locking rollers to stop the elevator plummeting to the ground in the case of an accident.

These days, elevators are a lot more complex. They have mechanical systems to manage the weight of the lift car and its passengers, along with electronics and safety components to operate the whole thing smoothly and automatically. There are two main options for modern elevators:

- **Hydraulic lifts** use a fluid-driven piston to push an elevator up from below. When the elevator is at the bottom floor, the entire piston mechanism needs to be housed underneath the building. You can imagine, then, that for a tall building, you'd need to dig pretty deep to accommodate a hydraulic lift.

- **Cabled lifts** use steel ropes, pulleys and a counterweight to raise and lower the lift. Think about old-fashioned weighing scales with one pan full and the other empty. To raise the full one, you'd need to push down rather hard on the empty pan. But in a perfectly balanced scale, with the same weight in each pan, either one could be raised with very little effort. This is how a counterweighted cabled lift works – it really only needs to overcome friction in order to raise or lower the car.

For our skyscraper, let's go with a cabled elevator. Luckily the building is short enough for a single elevator car to travel from the ground floor to the very top. For supertalls, this can't happen. Steel cables are heavy, and the longer the elevator shaft, the more cable you need, and the heavier and more impractical it gets. So for most skyscrapers you get one lift from the ground floor up to around 500m (1,640ft), and a secondary elevator to continue your journey to the top. However, a technology being developed by a Finnish cable manufacturer promises to let our lifts go further. In 2013, a high-strength carbon fibre cable, dubbed 'ultrarope', was launched, and by 2015 it was being installed in supertall structures in Jeddah and Auckland. Carbon fibre is exactly what it sounds like – a fibre made with thin strands of carbon (even thinner than a human hair). Like other fibres, it can be woven into a cloth, and if coated with a resin or plastic, it can easily be made into any shape. It's heavily used in Formula One cars and in a range of aerospace applications. But the main thing to know about carbon fibre at this stage is that it is incredibly strong and lightweight. It outperforms steel in terms of tensile strength, despite being 4.5 times less dense.

The ultrarope cable (which actually looks more like a belt) consists of bundles of carbon fibre embedded in a strong adhesive resin called epoxy. Its mechanical properties mean that the length of a lift cable could, in theory at least, be doubled, from 500m (as in the Burj Khalifa), to more than 1,000m (3,280ft). Daniel Safarik from the CTBUH believes that ultrarope is more than simply a significant move forward for the lift industry – he thinks it will enable the construction of a new generation of buildings. Speaking to me, Daniel said, 'It eliminates the need for transfers in order to reach 1,000 metres. If we consider elevators as the only limiting factor when it comes to height, then this tech might conceivably help us create buildings that reach up to 2km into the sky.'

While I might not quite go that far (I hope by now, we've seen how complicated it is to construct supertalls), this development is definitely a big deal. Elevator systems may be made to carry people or goods, but they also have to carry themselves (just like buildings). The materials in the car, the cable, the safety system and the electronics all need to be moved up and down in a controlled manner. Decreasing the weight of cable over the long distances travelled by skyscraper elevators would make everything easier to move, requiring less energy. A quick calculation would suggest that in our 380m (1,247ft) skyscraper, over 40 tonnes of material could be saved in rope weight alone.

So, now we have a lightweight elevator system, but how do we control it? In large buildings, an electronic control system directs the cars to the correct floors using a mathematical program called (unsurprisingly) an elevator algorithm. This offers the car three options: (1) continue travelling in the same direction while there are remaining requests, (2) if there are no further requests in that direction, then stop and become idle, or (3) change direction if there are requests in the opposite direction. Simple as this is, the elevator algorithm ensures that large numbers of people are moved up and down in the quickest, most efficient way possible. These days, more complex systems take the algorithm one step further, allowing some lifts to stop only at specific floors, reducing travel time.

There is another way to ensure we spend less time in an elevator – we can increase the speed of transport. The current fastest elevator in the world, which can be found in the Guangzhou CTF Finance Centre, rockets skyward at 72kph (45mph).* This is 24kph (15mph) faster than New York City's driving speed limit, and about twice as fast as an average skyscraper elevator. Speedy! Its manufacturer, Hitachi, says it is able to reach these speeds because the car is ultra-light and moves using clever control software and a powerful slim magnet motor (we'll come back to how magnets can move stuff in Chapter 2). Most elevators descend at just two-thirds of their maximum speed. That limit is set by those of us inside the car – any faster and our ears would struggle with the pressure change. But this latest generation of high-speed elevator cars can adjust their air pressure to minimise passenger ear discomfort. Other lift makers are developing speed-related innovations too, including twin elevators where two independent cars operate in the same shaft, and elevators that don't need a counterweight, saving valuable space in the process.

Our skyscraper is coming together nicely now, but there's one vital thing we haven't discussed yet...

Breeze

The single biggest consideration when designing skyscrapers is the wind. A building acts like a giant sail, so several engineering solutions are used to minimise any swaying and tilting. This is for more than just structural reasons. Humans are very sensitive to motion and, in a badly designed structure, can actually suffer from motion sickness during high winds. But perhaps surprisingly, making skyscrapers aerodynamic or wind-resistant is not the only possible option. In my chat with Bill Baker, I found out that some of the cleverest skyscraper design is about 'confusing the wind'. For him, when designing a skyscraper, you're 'mostly thinking about how to keep the structure from swaying; how to keep it stiff.

---

* This is equivalent to rising 20m (66ft) every second, travelling from the ground floor to the 95th in just 43 seconds.

Sometimes you might also have to consider seismic activity, but even in an area with moderate seismic activity, wind usually dominates designs.'

The thing is, for some shapes, such as a cylinder, the wind won't just flow by unimpeded – it forms areas of turbulence around the cylinder called vortices. If you've ever seen a streetlamp oscillating in high winds, you're already familiar with this kind of turbulence. If you scale that up to the height of a skyscraper, you have even more to worry about. While most structures can withstand some random wind buffeting, the issue with vortices is that they are not random – they occur again and again in a regular, repeating pattern (this is why they're called a *periodic* force). In the image below, you'll see a cylinder (black circle) followed by a series of blobs. Each of these is a wind-induced vortex, and together, they produce a kind of rhythm that can push the building from side to side. This phenomenon is called vortex shedding.

Now, before we go any further, we need to talk about resonant frequency. Picture yourself on a playground swing. In fact, if there's one in your neighbourhood, go and sit on it, in the name of science. We all know that in order to go higher on a swing, we need to kick our feet at *just* the right time. If we mistime our kick, the swing loses its smoothness and height. You may not realise it, but what you're doing is instinctively matching the natural (resonant) frequency of the swing. When you kick at the right moment, you cause resonance in the swing, and this amplifies the swinging motion, making you go higher. While this is a positive thing

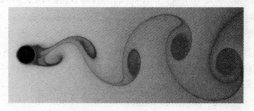

Figure 1.2   When wind hits a poorly-designed building, regular blobs of turbulence can build up behind that building. (Credit: Tim Colonius, CalTech).

for those of us who enjoy a visit to the playground, it can be a real problem for building design.

All skyscrapers have a resonant frequency – a speed at which they 'like' to wobble. If the wind on any given day produces vortices that match the natural frequency of the building (if it kicks perfectly in time), the building will begin to sway. As the vortices continue to regularly shed, the building's swaying motion will be amplified. If left unchecked, it could eventually cause catastrophic damage. Thankfully, all skyscrapers are designed with vortex shedding in mind – their shape is only partly to do with making them look gorgeous or interesting. Mostly, it's about stopping these vortices from forming in the first place. When I interviewed Baker about the Burj Khalifa, we spent a lot of time discussing the design considerations that were led by the wind. The building's triple-winged footprint truly confuses the wind because it effectively points in six directions at once, and breaks up any periodic wind patterns. Without those vortices forming, we won't see potentially damaging resonance.

The issue of vortex shedding does get more complicated though, because wind speed varies with height. And when we're talking about structures that reach over 800m (2,600ft) into the atmosphere, this effect is huge. Generally, the taller you go, the higher the wind speed (or velocity, v). Even more dramatically, the pressure that the wind applies to a skyscraper increases with its velocity squared ($v \times v$). So, as our building grows, it needs to contend with higher wind stress. This stress can cause the very top of even the most carefully designed skyscraper to move. Jason Garber, a wind-engineering specialist at Rowan Williams Davies and Irwin has said, 'For any tall building on a windy day, the amount of motion you'd expect is on the order of $1/200$ to $1/500$ times its height.' For the Burj Khalifa, this translates to the top moving by 2 to 4 metres (6.5 to 13 feet).

To counteract some of this height effect, most skyscrapers are narrower at the top than the base. But the Burj is a bit different from skyscrapers such as the Shard in London. Instead of using sloping sides, it narrows by a series of so-called setbacks – the building's height is separated into 26 differently shaped sections, each one narrower and in a different position from the one below it. This design gradually decreases the width of the tower, and it has a major added bonus. Going back to the swing analogy, these 26 setbacks effectively act as 26 legs, all kicking against the wind at different times, minimising the risk of vortex formation. Of course, this didn't come as a surprise to Baker and his colleagues. Long before a single pile was driven into the Dubai desert, the Burj Khalifa had been extensively modelled using a type of maths called computational fluid dynamics. As well as a lot of calculations, a physical model of the Burj was produced in miniature form, filled with sensors, and placed in a simulated variable wind environment.

This analysis did lead to some surprises. The initial designs for the building had the setbacks arranged in an anti-clockwise spiral, but modelling suggested that the building would be even more stable if the sections were instead arranged clockwise. And while the model showed that the building's three wings cut through the air as the prow of a ship cuts through water, vortices were building up on one side. But the team realised that when the whole structure was turned to face in a slightly different direction, those potentially damaging wind loads disappeared. All of the modelling, wind-tunnel testing and clever structural design means that the Burj Khalifa is the most stable skyscraper in the world.

Aerodynamics is just one way to help skyscrapers resist wind stress. Another option has been used to great effect in Taiwan. The Taipei 101 Tower houses a secret between its 88th and 92nd floors. Known as a tuned mass damper (TMD), this giant gold-coloured steel pendulum is the key to keeping the building upright. To understand how it works, we need to think about forces. Some of you might be familiar with the equation: $F = ma$. What it means is that when an external force, $\mathbf{F}$, is applied to a system with mass, $\mathbf{m}$, there will be an acceleration, $\mathbf{a}$.* We already know that wind can

---

* Technically, acceleration is the rate at which an object changes its speed. In equation terms, acceleration = change in speed ÷ time taken.

act as this external force, causing unwanted movement of the structure.

Tuned mass dampers counteract excessive motion in the building by 'leaning into' the force. Taipei's TMD is, in every sense, massive – it weighs in at 730 tonnes, but that represents less than half of 1 per cent of the building's mass.* The secret of how such a (relatively) small mass can stop a much larger structure from swaying comes in the 'tuned' bit of the name – it's all about its frequency. TMDs tend to sit near the top of a skyscraper, where the acceleration is greatest, so as soon as the building begins to sway in one direction, it is set into motion, via a series of springs, *in the opposite direction*. If the frequency of the TMD's motion exactly opposes that of the building's sway, the building will gradually return to a near stationary position. Tuned mass dampers can be produced in many sizes and shapes, and can be 'tuned' to withstand incredibly high wind speeds (216kph or 134mph in the case of Taipei 101). They are pretty damn impressive. And Taipei 101's TMD protects against more than just wind stress. The tower sits only 200m (656ft) from a major fault line, but thanks to its design, it managed to withstand a major 6.8 magnitude earthquake that struck during its construction in 2002.†

So, I think we're there with our skyscraper. We've learned about everything from foundations and concrete to elevators and wind, and even managed to take inspiration from some interesting people. The next step for us is a little different – our first foray into the future. How do you think our skylines will change as urban technology continues to develop?

---

* Mass and weight are not equivalent. Mass is a measure of the amount of stuff that makes up an object. Weight is the gravitational force that acts on that mass. The mass of something is constant everywhere in the universe, whereas the weight depends on gravity acting on it. So, stuff that feels 'heavy' in Earth's gravity will feel lighter on the moon, where gravity is weaker.

† However, the earthquake destroyed a crane on the roof of the building, tragically killing five people.

Tomorrow

Everyone I spoke to for this chapter agreed that skyscrapers will continue to dominate our cities. And the reason is simple: density. The Earth's population is still increasing, as is the trend towards living in cities. Figures from 2014 show that 54 per cent of the world's population (that's 3.9 billion people) live in urban areas, with the UN predicting a further increase by 2050. Urban sprawl and suburban living can accommodate some of these additional city residents, but for most experts, taller buildings are the logical choice. Ron Slade, from the team who developed the structural concept for the Shard, told me, 'We're starting to see both many more tall buildings, alongside a general height increase – so more supertalls too.' Slade also pointed out that 'a single skyscraper doesn't tend to be alone for long', a sentiment echoed by Bill Baker, who said, 'As the number of tall buildings in an area increase, we see some order appearing – an iconic skyscraper can help to form and organise a new city.'

In addition, most commentators expect to see a changing city structure. Baker suggests that in the future, we'll see fewer cities split into a central business district and surrounding suburbs. 'Low-density city sprawls are inefficient. Too often, people live in residential districts, and then travel into the centre for work. This offers a huge challenge to both urban planners and those managing power demand.' For him, the solution would be to 'have cities with multi-use districts that offer both places to work and places to live'. Related to this is another growing trend noted by numerous structural engineers: multi-utility. Single structures can provide a combination of functions – offices, apartments, shops, leisure centres, enclosed gardens… the combinations are endless.

The take-away message from my many skyscraper chats was that tomorrow's cities will be densely packed, and frankly that thought made me feel a little claustrophobic. Surely there are ways to make tomorrow's cities nicer places to be? Can science, engineering and technology play a role in creating future metropolises that work for all of us?

Materials

First, let's look at what tomorrow's skyscrapers might be made from. The first material might seem a little… antique. 'Timber!' exclaimed Bill Baker when I asked for the next big thing in skyscrapers. Baker and his firm recently revisited the design of a 42-storey concrete building and found that they could reliably rebuild it using timber. With tall wooden-framed buildings in Paris, London and Stockholm planned in the coming years, it seems that we may see a return to 'natural' materials. However, timber will always have limited applications. Baker says that, 'In reality, most tall buildings are around 30 storeys tall – and this is a sweet spot in terms of timber's properties.' Beyond that, we'll still need more heavy-duty, environmentally unfriendly solutions; but there are ways in which skyscraper construction will be made a little 'cleaner'. Across the world, teams are looking into cement substitutes, with many emphasising the use of waste products, such as flyash. A by-product of the combustion of coal, normally collected in the filters of power plants, flyash is being tested as an ingredient for cement. Its small particle size could offer a way to create dense, high-strength concrete, but it's still dependent on the fossil-fuel industry, so perhaps not a long-term option. Thankfully, there are lots of other substitutes for concrete on their way, but we'll have to wait until Chapter 4 to meet them. Sorry!

In the meantime, let's talk about glass, another exciting area for tomorrow's skyscrapers. Windows are an important factor in a building's energy rating, and improving their efficiency while keeping them transparent is an ongoing challenge. In the UK alone, up to 40 per cent of the nation's total energy bill comes from lighting, heating and cooling buildings, so even a small change in window technology could have a significant effect in reducing energy consumption. In late 2015, a group of Chinese researchers reported on their development of a self-cleaning, energy-saving and anti-fogging window. This new glass is effectively a sandwich of titanium dioxide (TiO_2) and vanadium oxide (VO_2). While you may never have heard of titanium dioxide, it is the superstar of pollution-guzzling materials – it is a photocatalyst, which means that when it's exposed to sunlight (more specifically, to ultraviolet light), it can speed up chemical reactions, without being used up itself. In this new glass, the TiO_2 surface breaks down any dirt that lands on it. The VO_2 has a different role – it can block infrared light (the light given off by hot objects). This means that it can be used to produce windows that don't let heat escape.

In 2016, researchers at University College London announced the development of their own self-cleaning, energy-efficient window. They used VO_2 to minimise heat loss too, but the cleaning process was very different. Their glass was covered in tiny structures 40 times smaller than a red blood cell. These bumps stops water from clinging to the surface, so it rolls off, taking any dirt or dust with it. In both cases, the result is an insulating window that cleans itself… and wouldn't we all love to see those on the skyscrapers of tomorrow?

Smart glass is another material we may see more of in future. As our cities become more densely packed, the need for privacy is likely to reach new levels, but who wants to close their blinds all the time? Electrochromic glass may be the answer to that – thanks to a layer of charged particles of lithium (called ions), it changes its appearance when activated by a small electric voltage. The voltage causes the lithium ions to leap from the inner surface of the glass to the outer surface, where they scatter sunlight and turn the glass opaque. When the voltage is reversed, the lithium ions move back to the inner surface, leaving the glass transparent once more. The amount of electricity needed to make the switch is tiny – the whole system can be powered by 5V (less than that supplied by 4 x AA batteries), so energy-efficiency-wise, these windows make a lot of sense. But they are still very expensive.

Thankfully, there are lower-cost, passive systems that don't rely on electricity. Instead, they respond to environmental changes, such as light levels (photochromic), or temperature (thermochromic). If you've ever used adaptive sunglasses, you'll know that photochromic materials darken when you walk out into the sun. Thermochromic windows are slightly different – they change colour as the day warms and cools, while remaining transparent enough to see through. These properties could offer variable tinting windows without the need for additional power. No matter whether we choose active or passive tech, smart windows would do more than just look awesome, they'd allow us to fine-tune the levels of light entering our homes and offices, and as a by-product, lessen the need for air conditioning. As far as I'm concerned, it's a win–win.

Another thriving technology that will influence how we build tomorrow is 3D printing. Also called additive manufacturing, 3D printing has become a bit of a catch-all term for anything that produces a three-dimensional object. In general though, it's the name given to a computer-controlled process that builds up a material, layer by layer. Although 3D printing has long been used in industry, these days we're more familiar with seeing tiny 3D-printed busts of Yoda than anything to do with cities. But, when scaled up, 3D printing could be used to build walls, and even houses. In fact, in 2015, a Chinese construction firm announced that they'd built a villa and a five-storey apartment block entirely from 3D-printed concrete (more on this in Chapter 4). And a Dutch architecture firm collaborated with a printer manufacturer to develop a large-scale system that prints a bioplastic made from 80 per cent plant oils. Called the KamerMaker, this 6m (20ft)-tall printer creates hollow shapes that can be assembled to form walls. The fact that they're hollow means that services (such as wiring and plumbing) can be installed easily, and the cavity can then be filled with other materials to add structural strength and insulation. However, the process is still slow and very costly, so it is unlikely to replace traditionally built houses any time soon. Still, definitely something for the far distant future.

Function

For many of my friends, I've long been that person-you-go-to-with-every-random-question-you-can-think-of, but once

I started working on *SATC*, that situation escalated. So on occasion through the book, you'll see me answer some of the most common questions I've been asked, thus immortalising them forever in print. So let's hope my answers are correct, eh?! First, **why do most skyscrapers have revolving doors?**

I'd like to think that it's at least partly because they are much more fun than your average door. But there is a somewhat scientific reason for their widespread use. The stack effect is a common problem in most tall buildings, and it's all to do with what hot air does (clue: it rises).[*] In cold weather, a skyscraper's heaters tend to be turned on, warming the air inside the building. This hot air rises to the top of the structure, leaving a vacuum on the lower floors. If the building had a regular 'swinging' door, opening it would cause cold air to be sucked in to fill the vacuum, thus creating paper-blowing, skirt-lifting winds throughout the lobby. In warmer weather, cold air-conditioned air sinks to the bottom of the building and rushes out when swinging doors open. The design of a revolving door means that it is always 'closed'. This prevents air from being sucked into or pushed out of the building, minimising internal whirlpools and keeping energy costs down. If unchecked, temperature variations caused by the stack effect can cause structural problems too, so rotating doors really are rather useful.

But for the skyscrapers of tomorrow, shouldn't we be able to make use of the stack effect? Well, a very ambitious plan in south-east China is hoping to do just that. If the project goes ahead, the Phoenix Towers in Wuhan will consist of two sci-fi-looking structures, set over a lake and marshland on the edge of the city. One of the towers will be, in effect, a thermal chimney. Solar concentrators (still to come in Chapter 2) will heat the air at the top of the hollow structure, and thanks to the stack effect, it will draw cool air through the building – a sort of natural air conditioning that can be used

---

[*] It does this because the atoms in hot gas are more dispersed (and so, less dense) than those in cooler gas.

in its rooms and offices. Lead architect Laurie Chetwood says that this 'will provide passive cooling throughout the lower, habitable floors of the building, effectively free of charge'. The taller of the two towers is planned to reach 1km (3,280ft) into the atmosphere, which would make it the tallest man-made structure in the world. 'But there is quite a lot in the height that is not just showcasing,' said Chetwood; 'a lot of it is to do with the environmental aspects of the structure.' This building may never make it off the planning pages, but there are some innovative ideas in there that are likely to feature again and again in our future cities. Watch this space!

Buildings that combine form with function will become increasingly popular in urban environments, but they come with their own challenges. Construction projects are always huge, complex operations that can involve thousands of people working together. While adding new high-performance materials or energy-saving technologies might seem like a no-brainer, the reality is that anything that adds risk to a project needs to be handled very carefully. And here, computers have a growing role. I'm sure some of you will have seen 3D renderings of buildings, which allow you to explore a digital version of the structure. These come under the heading of Building Information Modelling (BIM) and they've been around since the 1980s. But they're finally beginning to get a major upgrade.

In cities across the world, software is being used not only to see how a building will look, but also to plan how to put it all together, from individual internet cables through to the final pane of glass. By including details such as the mechanical strength of the concrete or the project timeline and budget, architects and engineers are adding a scientific slant to the construction process. Some engineering companies are taking digital construction one giant leap further. For example, Bechtel has developed a virtual reality system that allows people to take a virtual walk around a site. Longer term, they're looking to use this at the construction stage too, allowing workers on-site to visualise the final structure as they work on it. Part of this may include real-time environmental data and survey information taken from drones. It looks as if the future of construction may well be digital.

Green

I hail from Ireland, the Emerald Isle, so I understand the power of a beautiful green space to make you feel relaxed. But excluding public parks, greenery is not something cities are generally famed for. I spoke to Dan Safarik from CTBUH about a particularly exciting trend – the greenification of skyscrapers. The word 'green' is bandied around a lot these days, but in this case, it's all about plants. Living walls, also known as vertical gardens, have started to spring up in cities everywhere, and are widely hailed as both architecturally pleasing and environmentally friendly. Hmmm. Patrick Blanc is synonymous with these structures, but because he is a botanist rather than an architect, plant science really is central to what he does.* He insists on using 'the right plants in the right place', meaning only those native to the local environment, and those that grow naturally without soil. In addition, when designing his living walls, Blanc layers each plant type according to their light, water and nutrient needs. So, there is a lot of science behind *these* living walls – they are truly sympathetic to their surroundings and will make a lot of sense in the cities of tomorrow.

Unfortunately, the same cannot be said for many of the other vertical gardens you might see in your city today. Many are there simply to provide a particular aesthetic – they *look* green. These walls are incredibly complex to install and maintain, and they're based on costly, water-intensive irrigation systems and heavy, soil-filled grids. They also include non-native plants in the mix, which then need extra nutrients to grow and prosper. Yes, plants absorb carbon dioxide, and yes, they look beautiful,

---

* The inventor of the Green Wall system is Stanley Hart White, a Professor of Landscape Architecture who patented his system in 1938.

but the reality is that, ecologically, many living walls make no sense at all. And in terms of the building's design, they add only two things – a huge, ongoing financial burden and a lot more weight to support.

For Safarik, one project that is pointing the green way is Central Park, in Sydney, Australia. This mixed-use development in the heart of the city will boast the world's tallest living wall, designed by Patrick Blanc. The residential building also has a rather odd-looking platform jutting out of the 28th floor. In fact, this is a heliostat – a bank of fixed and motorised mirrors that can move to direct sunlight down onto the surrounding gardens and terraces, so that the building's 'shadow' is minimised. At night, the heliostat turns into an LED light show.

Of course, all of this extra stuff requires power, and so this Sydney development has its own tri-generation energy plant, run on natural gas. Tri-generation is often called combined heat, power and cooling (CHPC). In this system, electricity and heat are simultaneously produced. The third bit of the tri-generation triangle (the cooling) involves redirecting some of the hot water to an absorption chiller. These have no moving parts (unlike the pump used in your fridge) – instead they utilise the fact that every liquid evaporates at a different temperature. By mixing and separating liquids throughout, absorption chillers can move heat energy around, leaving cooled liquids behind. According to the design engineers WSP, Central Park's 2-megawatt plant could 'reduce greenhouse gas emissions by as much as 190,000 tonnes over the 25-year design life of the plant'. In practical terms, the WSP figures suggest that this would have the same effect as 'taking 2,500 cars off the roads every year for 25 years'.[*]

Not every greenification option needs to be super high-tech. One material that I think we're going to hear a lot more about is bioreceptive concrete. We've all seen dark spots of fungi or moss growing on buildings. Well, this material, developed by

---

[*] There is nothing I could find to independently confirm this claim, so maybe just take it with a pinch of salt.

Dr Sandra Manso Blanco during her PhD, emphasises that effect to create buildings unique to their environment. The concrete acts as a support structure for the growth of naturally occurring microorganisms; moss, fungi and lichens can thrive on its surface without the need for expensive irrigation systems. The technology is based on a multi-layered approach. First is a layer of normal concrete, which is topped by a waterproofing layer. Then comes the thin layer of bioreceptive concrete, followed by a still-to-be-perfected 'inverse waterproof layer' – it lets water in, but not out. As we discussed earlier, water has a very important role in curing concrete, so I was curious to find out how a concrete that keeps on trapping water could be structurally sound. 'The main way is to keep the two layers of concrete completely separate,' said Manso; 'only the inner layer is structurally important to the building, and only the outer layer encourages growth.'

In addition, Manso and her team are trialling a novel magnesium-phosphate cement that, unlike Portland cement, doesn't need to cure. Through adjusting the concrete's chemical composition and pH, as well as its porosity and roughness, Manso has perfected her bioreceptive concrete. The team is now working with three companies – ESCOFET 1886 (Spain), Manini Prefabbricati (Italy) and BASF (Spain and Italy) – to scale up production. I must admit that I love this project – it takes a natural process and uses materials chemistry to speed it up. On top of that, it will produce buildings that change as they 'breathe in' their city's air throughout the seasons. When I asked Manso if she'd seen any difference between trials in Belgium and Spain, she laughed and said, 'In Ghent we found lots of yeast – a particular strain that's widely used in beer! So these panels make buildings that are truly local.'

Clean

And that brings us nicely on to air quality. As the trend towards city living accelerates, tomorrow's cities will need to find ways to minimise these pollutants to protect the health of their citizens. Living walls and rooftop gardens can have a role to play in air quality, but they can't be used everywhere, so we'll need to look at other options, and soon. In 2014, the World Health Organisation (WHO) announced that in a survey of 1,600 metropolitan areas, nearly 90 per cent had air quality considered 'unsafe'. Industries, households and vehicles all produce complex mixtures of air pollutants, including oxides of sulphur and nitrogen (which lead to acid rain and smog), carbon monoxide and volatile organic compounds (greenhouse gases). While all of these pollutants can have an effect on health, perhaps surprisingly, the most dangerous pollutant to humans is fine particulate matter. Particles below 10µm (0.001cm) in diameter can cause respiratory problems, and right now, cities such as Peshawar in Pakistan show levels of these particles 27 times higher than the safe limit defined by WHO. An obvious solution is to regulate the processes that produce pollutants, and it is one that many governments have implemented. But there are also technologies that could be used within our cities to further clean the air.

What about some titanium dioxide poetry? In 2014, a collaboration between a scientist and a poet at the University of Sheffield in the UK resulted in a 10 x 20m (33 x 66ft) poster of a poem called 'In Praise of Air', coated with an invisible layer of titanium dioxide. For Professor Tony Ryan, who worked on the project, TiO_2 was an obvious choice – in the presence of light and oxygen, it helps to add oxygen to the nitrogen oxides (NO_x) in urban air and transforms them into inert compounds that can be washed away. In short, the billboard cleans air up. Ryan calculated that his billboard, placed by a busy road, absorbs the NO_x produced by about 20 cars each day. Importantly, the TiO_2 in the coating is just the catalyst in the reaction. This means that it 'shakes up' the oxygen atoms, but it isn't actually used up. So, for as long as the coated poster is in use and isn't too weatherworn, it will continue to purify the air. The coating adds a cost to the advertisement (Ryan estimates an additional £100 for this billboard), but if every poster in a city had this coating, we would see a noticeable improvement in the area's air quality.

There are other ways that the photocatalytic properties of titanium dioxide can be retro-fitted to existing structures. A group of students from the University of California Riverside recently coated a series of commercial clay roof tiles with the material. They found that the coated tiles removed huge quantities of nitrogen oxides from the surrounding air in a closed chamber. A thick coating was found to remove almost all of the pollutants, but even a very thin coating, costing just $5 per roof, removed 88 per cent of the NO_x in the air. According to their results, a typical residential roof would break down the same amount of nitrogen oxides as an average family car produces by driving 18,000km (over 11,000 miles). Impressive. For the Manuel Gea González Hospital in Mexico City, TiO_2 came in the form of a decorative facade that was used to give the building an ultra-clean facelift in 2013. Not only do its honeycomb panels look gorgeous, but their complex geometry is based in science too, because it's all about surface area. We know that a flat panel coated in TiO_2 can help to clean the air, but by increasing the surface area – that is, coating a more complex shape full of angles, nooks and crannies – many more reactions can take place. This speeds up the whole process and means that even more air can be cleaned. The addition of the facade to the Mexican hospital increased its surface area by 200 per cent, hugely improving its ability to destroy NO_x. According to Elegant Embellishments' Daniel Schwaag, who developed the modules, 'This facade could neutralise roughly the same amount of smog produced each day by about 1,000 vehicles in the highly polluted city.'

Ageing

An often forgotten consideration for tomorrow's megacities is our ever-ageing population. According to WHO, for the first time in human history there will soon be more people on Earth aged 65 and over than there are children under the age of 5. While increasing life expectancy should be appreciated as the remarkable achievement it is, ageing populations bring with them a range of new questions, challenges and opportunities that will shape our urban homes.

To learn more, I spoke to Professor Ruth Finkelstein from the Columbia Ageing Center. She started off by saying that for city planning, 'The winning strategy would be to plan a city for all ages – age segregation isn't really good for anyone. It's less about *prioritising* older people, and more about *including* them.' According to Professor Finkelstein, there are a few key things that would make the future city more accessible to all. Open public spaces, parks and car-free retail centres would allow people to mix freely, and help avoid the social isolation for which cities are notorious. Technology has a big role to play too – widespread high-speed internet made freely available would help older 'digital migrants' to learn how to use it. 'Cities like Manchester, Barcelona, New York, Rio de Janeiro and those in Quebec are all doing excellent work in this area. Really the ideal city of the future should help to bring people together rather than simply co-exist.' I feel as if we still have a long way to go on that front.

Harvest

It seems that we're looking to a future with many more high-tech, multi-use and supertall skyscrapers. They can even change their appearance and clean themselves! Our cities, although dense, could be greener than ever, with living walls, tiles and billboards that can purify the air. Maybe the urban population won't just grow older – it'll grow wiser too. So what's left? Arguably the most important topic of all: energy.

The skyscrapers of tomorrow will have to do much more than just sit there, looking green and gorgeous – they'll also need to produce some of their own electricity, and they have many options. Things like wind turbines and solar cells are already making a name for themselves in today's skylines. We're going to delve into the details of these technologies in Chapter 2, but to prepare ourselves, we need to talk about the conservation of energy.

This phrase might be familiar – in short it means that energy can't just come from nothing, and it can never be destroyed. It simply transforms from one form to another. This is universally true, though most of the time, we are completely unaware of it. Let's take a portable radio as an example. The chemical energy in the batteries is first transformed into electrical energy and heat energy through the metal wires (and the warmed-up battery). The electricity causes the speakers to vibrate and they produce sound energy, allowing us to hear the music. The point is, the energy stored in the batteries doesn't disappear, it just changes into lots of other forms of energy, and some of it we don't want, but some we do. All electricity technologies do the same thing – they collect energy, be it solar, wind or thermal energy, and transform it into electricity. Switching between different forms of energy *doesn't break any laws of physics* (special shout-out to the man who cornered me with that accusation at a talk I gave on energy harvesting), but it will have a huge relevance to our future cities.

In his excellent book *Sustainable Energy – Without the Hot Air*, Professor David MacKay calculated that even in not-so-sunny Britain, the average raw power of sunshine that hits a square metre of south-facing roof is roughly 110watts/m^2. Closer to the equator, which receives the highest incidence of direct sunlight, this figure is substantially higher. This goes some way to explaining why photovoltaic cells (generally just called solar cells) have become so popular in sunny cities. But why aren't these cells already everywhere? If we could capture all of the available power from the sunlight that reaches us, we could run our lives on it!

The key word here is 'if'. Solar panels cannot transform all of the sun's energy into electricity. In reality, even the best manage less than a third, with most of the rest converted to (mostly useless) heat energy. So why should we bother? Frankly because if we don't, we're idiots. Here is an effectively infinite source of energy – even tapping into a small fraction of that makes sense, especially in the light of growing energy consumption in cities. And solar power isn't destined to remain an 'under-achiever'; as we'll soon discover, tomorrow's solar cells will be game-changers.

Wind turbines have a similar issue – they lose a lot of their energy to noise and vibrations, which makes integrating them into a building rather challenging. Some have managed it though. In 2008, the Bahrain World Trade Centre became the first skyscraper to integrate wind turbines into its design, with three huge, 29m (95ft)-diameter wind turbines on the connecting bridges between the structure's two towers. The sail-shaped buildings had been designed to funnel wind through the gap, and to the turbines. And they worked! Bert Blocken, Professor of Building Physics at Eindhoven's University of Technology, showed that they could have been so much better, though. Using a series of wind tunnel tests and computer simulations, Blocken showed that if the towers had been positioned slightly differently, the turbines would have produced 14 per cent more wind energy. Doh!

Despite that minor glitch, integrated turbines do seem to be increasingly popular in cities, and wind power is something we'll come back to in the following pages. For now, though, I think that's enough time travel. This chapter has been something of a whistle-stop tour of our skylines, past, present and future. But in meeting leading experts and exploring some of the technologies surrounding us, we've laid a lot of the groundwork for the rest of the book. Consider yourself initiated into the science of cities.

CHAPTER TWO
Switch

Now, it's all very well having a landscape full of skyscrapers, but that alone doesn't make a city. Since the year 2000, astronauts on board the International Space Station have treated us to remarkable images of the Earth. Key among them are the night-time shots of the world's cities, and from that unique vantage point, it's easy to see that they are more than a collection of buildings. A city is a vast, seemingly living, breathing organism, criss-crossed with countless veins and arteries – all different, all vital. From the highways and the rail tracks, to supplying water and disposing of waste, it's these networks that have shaped our cities and allow them to function; but do you know how they work? Throughout the next few chapters, we're going to look at these systems in detail, considering the science behind each and the engineering that made them possible.

I'll look to comedian (and all-round physics fan) Dara Ó Briain to get us started on the first topic: electricity. In one of his classic sketches, he imagined describing modern technology to the great minds of the Renaissance. Dara gets as far as 'It's connected to the wall' before giving up. By the time we reach the end of this chapter, we'll be able to cast off that cloak of naivety. Together, we'll uncover just how electricity makes its way to us, and where your city gets it from. We'll also learn about the amazing research that will transform the power grid of tomorrow.

Today

First, when I say that electricity has shaped our cities, I'm really talking about the modern city. Other factors such as geography and reliable access to water dominated the civilisations of the ancient world, but when the electric lamp first lit up a main street, the city as we know it was born. Electricity changed the world because it provided a way to 'pipe' energy around, and today, we are at the mercy of it. We've never produced or used more electricity than we do right now, and it's our urban hubs that demand it most of all. With more than half of us now living in cities, it seems reasonable that they produce about half of the world's waste. However, cities are also responsible for up to three-quarters of global energy use, and they emit around the same proportion of the world's greenhouse gases. Depending on what city you're in, your electricity might come from fossil fuels (oil, coal or gas) or renewable sources (wind or solar), or more likely a mixture of both. Exactly what we use it for may surprise you. But before we leap into all that, we need to cover some of the basics.

Electron

Think about a typical weekday morning. After sleeping through your alarm, you stumble towards the bright lights of the bathroom, and hop in the shower.* Then it's down to the kitchen for a coffee and a slightly burnt slice of toast. You set the washing machine on a timer, before grabbing your phone from its charger and running to catch your bus or train. OK, I'm generalising, but the vast majority of city dwellers couldn't get through their daily routine without using a lot of electricity.

But what exactly is it? Well, let's start by saying that electricity is not something humans invented. On a fundamental level, everything is electrical, and that's down to what's found in every atom in the universe: electrons. Cast your mind back to the picture of an atom presented in school books – a small, dense ball (containing protons and neutrons) in the centre, surrounded by shells of whizzing electrons.†

---

* Other bathroom activities will be covered in detail in Chapter 3. You have been warned.
† In reality, atoms don't look like this at all, but it's an image we all know, so it will do for now.

Electrons are tiny, but just *how tiny* is a matter of ongoing debate. If we were to enter fully into the debate, we would be dragged into the strange and murky world of quantum physics. Now, while I love to visit the quantum realm on occasion, it probably falls outside the remit of this book.* So let's remain at a safe distance for now, and say that depending on what definition you go for, electrons are between 10^{-15}m and 10^{-17}m in size. If we could line electrons up side by side, we'd fit around 10,000,000,000,000 of them inside a humble, flavour-giving dried peppercorn.† That number is a bit larger than the total number of stars in the Andromeda Galaxy, all squeezed into a space the width of a corn of peppery goodness. Electrons are *really* small.

More importantly (for this chapter at least) every electron carries a negative electric charge, and it's by ripping them away from their home atom and making them flow, that we get electricity. It might sound vicious, but electrons are very easy to steal. If you've ever given yourself an electric shock by touching a metal handle, you are an atomic home-wrecker. The jolt you feel is the rapid jump of electrons from you to the handle. When rain clouds rub up against the air around them, you get the same effect, but the outcome is more spectacular – lightning. All of these are forms of static electricity, and while its 'jumps of electrons' contain a huge amount of electrical energy, static is not nearly reliable enough to run our TVs on.

Instead, when most people talk about electricity, they're usually referring to current electricity. This is the flow of electrons that everything connected plug-to-socket relies

---

* If quantum physics is what you're after, my favourite guides include Chad Orzel's *How to Teach Quantum Physics to Your Dog*, Louisa Gilder's *The Age of Entanglement*, and anything by Richard Feynman.
† We can't do this because electrons are negatively charged, so would repel each other. Quantum physicists would argue that electrons aren't physical things at all – they're more like a cloud of probability… Look, I warned you it was strange.

on. To get electrons to flow, we need to provide a welcoming path. Metals like copper are used in electrical cables because they have 'free electrons' that move with only the slightest encouragement. This means an electric current can easily flow through copper, and the size of the electric current is measured in amperes or **amps** (symbol: A). Next, we need to give the electrons in our wire some motivation to get moving, and for this we use a mains supply or a battery. These provide the energy – also called potential difference – needed to kick-start the flow of current. The bigger the potential difference, the more current will flow. You may know this force better as **voltage**, and it's measured, unsurprisingly, in **volts** (symbol: V). If we combine these two, we get **power**, and this is measured in **watts** (symbol: W).

The path that current electricity flows through is called a circuit, and that's for good reason. If you stick a wire onto one terminal of a battery, nothing will happen. A current will only flow when there is a complete loop – when that same wire is connected across both sides of the battery. That alone would be pretty pointless, but if you add other objects into the circuit that can use that energy, you're onto a winner. This is how the mains supply in our homes and businesses works – when we plug in a device, it enters that circuit and draws power from the mains. Let's put a 60W lightbulb into a new lamp and plug it in. I'm afraid there is no standard 'mains voltage' in our cities – if you're in the UK, it is around 240V and in the US it is 120V.* To balance out the numbers, and to make sure that a 60W bulb shines as brightly in the US as in the UK, the size of the current must be different. In this case, in the UK, 0.26A would flow through the bulb, but in the US the current would be higher, at 0.5A. This straightforward relationship between power, current and voltage is pretty useful when understanding electricity.

---

* I'm not being UK/USA-ist. Pretty much every country in the world has a mains voltage of either 110–120V or 220–240V.

Consume

Now that we've familiarised ourselves with electricity and introduced a few of the key terms, it's time to start putting it into context. I'm not going to tell you how every appliance in your house works.* Our interest really lies outside the home, on and under the streets of your closest city. But there is one thing I want to address before we get back on track, and it's inspired by a question I've been asked countless times. In this land of volts and amps, **where does the kilowatt-hour (kWh) on the electricity bill come from?**

As well as linking voltage and current, power is also a measure of how much (electrical) energy a device uses every second. If we wanted to find out the total amount of electricity our TV used in one hour, we'd multiply its power (in watts) by the number of seconds in an hour (3,600). This number could get very big very quickly, so instead of dealing in watt-seconds, it gets scaled up to kilowatt-hours.† In short, this weird unit is just another way to define energy usage. Let's take an example – plasma TVs run on around 50W. In order to use 1kWh of electricity, the TV would need to be on continuously for 20 hours (boxset, anyone?). At the time of writing, this TV-watching marathon would cost about 10p in the UK (15c in the US), and possibly leave you sleep-deprived for a week. But hey, maybe it's worth it. The *real* cost of the electricity is in the generation stage, and there is more than one way to generate it, as we'll soon discover.

Of course, electricity powers much more than the typical home: it is the bloodstream of the modern city. Streetlamps and escalators, trams and trains were all made possible by the introduction of safe, cheap electricity into our historic cities. These days, cities munch through a remarkable amount of energy. According to data released by the UK's Department for Energy and Climate Change in 2012, London used

---

* But if you are keen to learn more on that, I highly recommend *Atoms Under the Floorboards* by Chris Woodford.
† Where 1kWh = 1,000Wh.

108,467,000,000kWh of energy in just one year.* This is around twice as much as Dubai consumes, and a little less than New York gets through, but by any metric, the numbers involved in energy consumption are vast. In all major cities, the bulk of it is used by its buildings. Koen Steemers, an architect from Cambridge University, suggests that a city's buildings may use *twice as much* electricity as its transport network!

Not all buildings are equal though, as researchers from Columbia University demonstrated in 2015. Working with the New York City Mayor's office, they mapped the energy consumption of the entire city. This involved collecting data from over 850,000 buildings, classifying them into different categories and doing a lot of maths. They produced an interactive map to illustrate it all, and I'd recommend having a play with it.† Their analysis showed that New York office buildings use most of their energy on basic electric applications – 'appliances, lighting, ventilation, and refrigeration'. For residential buildings, which make up the *majority* of NYC's building stock, the use was rather different – most of their energy goes into hot water production and space heating. New York is not the only city where a huge proportion of its energy budget goes towards heating. In fact, in 2012 the European Commission said, 'Space heating represents the single largest electricity end-use for consumers in the residential sector.' So where does all the power come from to keep us warm? Heating is rarely the job of the electricity grid – in New York and London, heating systems rely heavily on the burning of fossil fuels, but this is something both cities want to move away from. Elsewhere, especially in mainland Europe, they've taken a rather different (and very cool) approach, and we'll talk about it soon.

Right, enough of tempting you with things yet to come. Look around you. Can you see an electric socket anywhere?

---

* This includes both electrical energy and the energy for heating, sourced from the natural gas network.
† You'll find it listed in the references section at the end of *SATC*.

That is your access point to the city's electricity network, and it can pump out rather a lethal amount of power, so please don't do anything silly, like sticking a fork in it. We're going to work our way out from that modest little hole-in-the-wall, right through the distributed network of cables that we call the grid, and back to the myriad of electricity generating stations that our cities depend on. First question, what shape is your socket?

There are approximately 15 different sockets currently in use across the world – a hangover from the days before electricity (and people) went global. But they all have one job to do, and that's to deliver the correct voltage to your gadgets. Overhead transmission lines regularly carry electricity at voltages between 132,000V and 755,000V, but when it gets to us, it's way down at around 120 or 240V. So, what changes along the way? The answer is that the voltage has been transformed. Some transformers star in over-the-top Hollywood blockbusters, but others have a much greater purpose. In our electricity network, transformers are the gate-keepers.

Here's a curious fact about electricity: when a rapidly changing (or alternating) current moves through a wire, it produces a magnetic field pattern around it.* The same is true in reverse – put a metal wire in a changing magnetic field, and it produces an electric current. This phenomenon is called electromagnetic induction and transformers put it to good use. Take a square block of iron and cut out the centre. Now you have a core for your transformer. Carefully wrap a wire a few times around one of the sides of the iron core, leaving the two ends free. Do the same on the opposite side. Connect one of the two wire coils to a voltage source, and the current that flows through it induces a magnetic field in the iron. When that magnetic field reaches the other wire coil, it induces an electric current in it. So what? Well, the clever bit is that the number of turns in the wire coil defines how big the current will be. We can use induction to make a

---

* We're going to talk about alternating current in a little while.

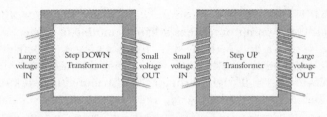

Figure 2.1 To use a transformer to reduce our voltage, we'd need to put more turns on the first coil than on the second coil. If we wanted to increase our voltage, our first coil should be smaller than our second coil.

voltage bigger or smaller, just by putting a different number of wire turns on each side of our transformer.

Repeatedly stepping voltages up and down might seem a bit odd, but it means that different bits of the city can receive different voltages. While 240V might be enough for domestic use, it's not much good for anything larger. The Madrid train service runs on 3,000V, and a factory might require 30,000V. Whatever voltage your city needs, you can design a transformer to step it down from the high voltages used on transmission lines. Transformers manage our supply, letting only what's needed enter or leave, and they're used in every city in the world.

Grid

So now that we've passed the gate, we enter the electricity grid, proper. One part underground and one part overground, the grid is a maze of interconnected cables and wires that connect us to our various electricity generators. In Vancouver, 90 per cent of electricity is generated from hydropower, but for Delhi, coal and gas make up the vast majority of its capacity. Most cities lie somewhere in between, with a combination of different sources – some that continuously generate electricity (e.g. nuclear) and others that produce it periodically (e.g. wind). The challenge of managing the complexity of the grid was summed up well by engineer Jamie Taylor when he said, 'The miracle of electricity generation and supply is its extraordinary balancing feat.' The age-old adage of supply versus demand is at the heart of grid engineering, and we'll come back to it again and again.

I should say that, at least on a local level, the electricity *grid* is more like a network of tree roots – these are called distribution lines, and in most built-up urban areas, they're buried underground. London's distribution network stretches over 30,000km (18,600 miles) – that's enough wire to wrap around the moon almost three times! The cables themselves are usually made from braided copper or aluminium wires, and they are covered in a tough plastic sheath to protect the metal from us, and us from the power they carry. If we follow one of these lines, we soon reach a rather important box. It turns out that the transformer just outside our house, the one that steps our voltage down so that we can safely plug in our gadgets, isn't working alone. In fact, it's part of a chain of step-down transformers, better known as substations, across the distribution network. Usually painted green or grey, these boxes transform huge quantities of electricity every day, so forgive them if they look a bit underwhelming.

The more visible part of the electricity grid is the transmission network. This begins with the substations, and includes thousands of kilometres of cables supported by poles and pylons. Its job is to transmit the electricity from wherever it has been generated to the distribution network. The cables here are much thicker than those found underground, and they're not coated – hence the need to keep them far out of reach. You might be wondering why we use such high voltages on transmission lines. Believe it or not, it's to save energy. We know from Chapter 1 that energy can be changed from one form to another. In the case of transmission lines, if we use a high current, some of the electrical energy is transformed into heat energy and lost into the air. But if we transmit electricity at a low current (and high voltage) we get much less heating in the lines, saving lots of energy. This brings me on to an important question. **How can birds sit on high voltage lines without being electrocuted?**

Admit it, you've seen birds lording it over you from their urban perches on the high wires. As with everything electricity-related, their survival is all about connections and very little to do with inherent bird powers. It turns out that animals are very conductive indeed – a good path for electricity to flow through (blame the fact that we're mostly water). If you hold a battery between your thumb and finger, you're effectively completing the circuit. But when a bird (assuming it is a wise bird) sits on a high-voltage line, it sits on just one, so there is no circuit. It's the equivalent of putting your finger on just one terminal of a battery – no circuit, no flow of current.

Now this is absolutely NOT to say that if you touched just one cable in a high-voltage line, you'd be fine. You'd be dead, so don't do it. Birds are OK because both of their feet are on the same wire – they touch nothing else. But if we tried to touch it, we'd need to be connected to the ground (via a ladder or a cherry picker). Our body would then give the electricity a fast track to complete the circuit, and it really, really wouldn't respond well to it. Equally, if a bird were unlucky enough to touch any of the other cables (or the pole) while standing on the wire, it too would be fried to a crisp. So, don't play with electricity, kids.

This important lesson was clearly not made available to one of the key figures in the history of electricity, who famously both began and ended his formal schooling at the age of eight. He was also almost entirely deaf by the time he was 12, but neither handicap appeared to slow him down. In his 84 years, Thomas Alva Edison amassed over 1,000 patents, including, famously, that for the incandescent light bulb.* For me though, this isn't his greatest invention. Edison really made his mark by opening the world's first integrated power plant and grid system. There is still a lot of debate over whether Edison first did this in New York or Brockton, Massachusetts – the historian Gerald Beals believes that Brockton came first but that it was downplayed to 'maintain and enhance media focus upon … the great metropolis of New York City'. Whatever the truth, we know that Edison opened two

---

* There's plenty to suggest that Edison didn't invent the lightbulb, though. A British inventor, Joseph Swan, had obtained a patent for it a year earlier, and took Edison to court for patent infringement. Swan won the battle, but given how synonymous Edison's name is with the invention today, he lost the war.

plants between 1882 and 1883, and that each one used a different wiring system: two wires for New York and three for Brockton.

So how did they compare? The two-wire grid used one enormous generator to ensure that the entire network received the same amount of current. In order to carry that current, the wires had to be very thick copper cables; a pricey option in the late 1800s. In contrast, the three-wire system made use of two smaller generators. This split up the current across three wires, allowing them to be much thinner and still safely deliver electricity to each streetlamp. While both systems were functional, the three-wire approach was described in an 1884 edition of the journal *Science* as 'highly ingenious', and its descendants form the basis of today's electricity grid. To many, this makes Edison the undisputed 'Grid Father'. I don't agree, but that's partly because, long ago, I lost my heart to a charismatic Serbian engineer who we'll meet in a minute. But one thing is for sure: although Edison didn't do it alone, he did have a huge role to play in making electricity generation and city-wide distribution practical, safe and cheap.* There was one thing he got wrong, and to understand it, we need to talk about how we generate electricity.

Generate

At the risk of repeating myself, it's important to remember that we can't create or destroy energy, but we can change it from one form to another. So, the key to an electricity supply is to find another type of useable energy, design a system to harvest it, and then turn it into electrical energy. Sounds easy!

Never underestimate how hard it is to generate electricity. Back in 2011, myself and some of my old colleagues from the National Physical Laboratory demonstrated this at an exhibition at the Royal Society in London. We put everyone from government ministers to national rugby players to the

---

* He was famed for patenting technologies he'd 'acquired' from collaborators and competitors alike. It was a British scientist called Joseph Hopkinson who first proposed the idea of the three-wire system. Edison bought the concept from him and made it a reality.

test on a dynamo bike. Without exception, the contestants were left sweaty and exhausted by the effort needed to continuously light the bike's headlamp. Most cities with cycle hire schemes use a similar technology in their bikes, and, if we scale it up massively, it's how city-sized power plants generate electricity too. It comes back to the same effect that transformers tap into – electromagnetic induction.

As a quick reminder, induction involves three things: magnetism, electric current and motion. Combine any two of them and you get the third. So if we want to generate electric current, we'll need a moving wire and a magnet. Pushing a wire in between the poles of a magnet just once will produce a very small 'blip' of current, which isn't helpful. Ideally, we would move it many times and very quickly to produce a reliable, continuous current. Electric generators achieve this using a long piece of wire, wrapped into a loop (like wrapping a bit of string around your fingers). The loop is then attached to a shaft that is spun at high speed inside a magnetic field: moving wire + magnet = current. Just as in the transformer, the more 'turns' you have in your wire loop, the larger the electric current you produce. Other ways to increase the output of your generator are to speed up the movement and/or to use a bigger magnet. The electric generators found in power plants (nuclear and fossil fuel), wind turbines and hydroelectric systems are all based on this principle, with a few tweaks. Our cities would be lost without them.

When these systems were first developed in the late 1800s, there were two schools of thought: one side was led by Edison and the other was headed by a young, handsome, ridiculously cool engineer called Nikola Tesla (honestly, if I could draw a big engineering heart around his name, I would). Dubbed 'The War of Currents', the battle between them was a surprisingly brutal affair that had implications across the globe. There are two ways that electric generators can produce current and it's all to do with direction. Picture a dial with a moving needle, like one on the dashboard of a car, only this one measures electric current. As a wire moves into a magnet, the needle swings in one direction, and when we pull it out, it swings in the opposite direction. So as our wire loop spins around inside the magnet, it naturally produces what's called alternating current (AC) – a current that constantly flips direction. Tesla and his army favoured this option for electricity generation and transmission. On the other side of the battlefield was Edison. He wanted to get away from this 'flipping' current, so he added a connection to the loop that meant the current was produced in one direction only. This was called direct current (DC), and it's also the way that batteries and solar cells supply electricity, so it's a rather useful option. But DC just wasn't suited to the long-distance transport of electricity, and transformers don't work with direct current. In the urban jungle, offering one voltage to all users was never going to catch on. Tesla and his contemporaries won the battle, and today, all cities are powered by AC electricity.

It turns out that many of the city-related questions I was asked by friends and family were about electricity. Besides the non-electrocuted birds and the electricity bills, the most common one was, **why do some transmission lines make a buzzing sound?** The answer is partly to do with the fact that, thanks to Tesla, we transmit electricity at high voltages. The voltages are so high that, sometimes, they can charge up the air close to the cable – the normally calm, relaxed nitrogen molecules are assaulted by the electric field, and have their electrons ripped off. This happens incredibly quickly because of how fast the current alternates. A city's generators may 'flip' 50 or 60 times every second – so fast, it looks blurry. So what you're hearing is effectively the 50-times-a-second assault of air molecules.

Anyway, what were we talking about? Oh yes, generators. In theory at least, the electromagnetic induction bit of the process doesn't produce greenhouse gases. What *does* is the step before that – getting the energy into the generator, in order to make its wire loop spin. For fossil fuel and nuclear plants, this energy is harvested in the form of heat. As we'll talk about in Chapter 5, there is a huge amount of chemical

energy stored in fossil fuels, and the best way to release that
energy is to burn it, which is what power plants do.* This
heat is then used to bring ultra-clean water to incredibly
high temperatures, upwards of 540°C (1,000°F), leaving
you with a constant supply of superheated, high-pressure
steam. The steam is then forced into a turbine – a collection
of thousands of tightly packed steel blades all arranged
around a shaft. The flow of steam past the blades causes the
turbine to rotate, which spins the wire coil on our
generator.

This burning–boiling–rotating process is incredibly
inefficient – only between a third and a half of the energy
stored in the original fuel is actually converted into electricity.
The rest of it is effectively wasted, and it's this that makes
fossil fuel plants so environmentally unfriendly. Burning
huge quantities of coal, oil and gas to boil water releases
tonnes of carbon into the atmosphere. As a result, the world
is warming up and our climate is changing – the data is
unequivocal on that.† Regardless, they are still the main
weapon of choice in the battle to produce energy. According
to UN Habitat, in 2012, the 'global energy supply consisted of
81.3 per cent fossil fuels, 9.7 per cent nuclear power, and only
9 per cent renewable energy sources (such as hydro, wind,
biomass and solar)'. This weighting is partly for historical
reasons, and partly due to cost. Although inefficient, fossil
fuel power plants still offer 'cheap' electricity because the
systems have been in place for a long time, and we've not yet
run out of fuel. It's very much a case of 'that's how we've
always done it'.

But there is a trend emerging. At municipal, national and
international levels, efforts are underway to decarbonise
the grid – to move away from relying on the burning of

---

* In nuclear power plants, the heat is not produced in this way. It
comes from the energy released by breaking atoms apart.
† If you're an outright climate change denier, I think it's safe to say
that this is not the book (or certainly, the chapter) for you. Or who
knows, maybe I'll change your mind!

carbon-rich fossil fuels, and towards those options that produce electricity without burning anything. Solar power and wind turbines are leading the way on this, and over the last decade they've become steadily cheaper. A report published by Bloomberg New Energy Finance in late 2015 showed that the trend may even be speeding up. They analysed the overall cost of generating electricity by different means, and calculated it in dollars per megawatt-hour ($/MWh). This figure is determined by looking at the whole lifetime of a power plant, from digging the foundations to its eventual decommissioning, to provide a level playing field for electricity competitors. The research found that in Germany, coal and gas are now more expensive than onshore wind – $106 and $118 versus $80/MWh – and the same is true in the UK. In China, coal remains king, coming in at just $44/MWh, but solar power is now cheaper than gas ($109 versus $113/MWh). In many cases, fossil fuels are gradually becoming more expensive, while renewables reduce in cost. The tide is turning for electricity generation, and this is beginning to change the way we design and manage our cities.

Heat

Now, before we launch into the physics behind renewable energy, we need to take a small but important detour. Not all of a city's energy comes from its electricity supply. Any figures that quote 'global energy supply' include both electricity consumption and the energy we use to warm our rooms and heat our water.

As we mentioned briefly right at the start, in cities, heating is one area still tied to the burning of fossil fuels, but not in some faraway power plant. Boilers found in offices and homes in almost every city work by burning natural gas to heat water. This is then pumped around the building and put to use in heating systems. In many of the world's urban centres, the natural gas network actually provides *more* energy than the electricity grid. Of course, closely related to heating is cooling, and this is also costly. According to the United Nations Environment Programme (UNEP), more than half of Kuwait City's annual energy bill is spent on air conditioning, and Dubai is similar. Urban heating and cooling is a huge problem for energy suppliers, and one that will become ever more challenging as our cities get larger and denser. In 2011, the provision of hot water, space heating and cooling were estimated to account for roughly half of global energy consumption in buildings. And no-one expects this proportion to drop any time soon.

But there may be a way out for cities, and it might already be found under the paving stones of your street. Called district energy, it's a system that pumps water (cold, hot or steaming) around an urban area in heavily insulated pipes. Large municipal buildings or individual communities can connect to the system, tapping into the water and using it in heating or cooling. There are two particularly clever things about district-wide systems: (1) they don't use fresh water, so it doesn't affect the drinking supply, and (2) they can make use of the wasted heat energy from our power plants, or can utilise alternative heat sources, such as the burning of non-degradable solid waste.

If you are in Copenhagen, Helsinki or Warsaw, you already rely on this network to supply most of your heating. In fact, many Danish neighbourhoods have entirely done away with their individual boilers. Instead, their hot water comes directly from their highly efficient communal district boilers. If you're in Vancouver, Izmir or Tokyo, you tap into a similar system for your cooling needs. Paris is the proud owner of Europe's largest district cooling network. Supplying water to cool down the city's hospitals, social housing and public buildings (such as the Louvre), it even makes use of water from the River Seine. More than 30 other major cities are investing in this technology too, so consider this a sneak peek into managing heat in the cities of the future.

Now that we've covered the basics of 'traditional' electricity generation and talked a bit about heat, it's time for us to move on to renewables. I've included just two in this section: wind and solar. Not only are these systems already a familiar sight in the urban landscape, but they're also the focus of some exciting research that will genuinely transform them in the years to come.[*]

Wind

Let's start with wind turbines, especially as everyone has an opinion on them. I must admit that I love how they look, but what really interests me is the engineering, physics and materials that make them possible. For example, there's a reason they're so tall – wind turbines want to capture the smoothest, highest-speed wind they can. Close to buildings and other structures, you get a lot of disturbed, turbulent air. Besides that, just the very act of having to flow over rough surfaces, be it a tarmac road or a rolling green hill, hugely reduces wind speed. There are a couple of different equations that quantify just how much of an impact height has on this, and both show that the higher you go, the faster the wind blows.[†]

This is important because the electric output of the wind turbine's generator is directly linked to the speed of the wind that turns its blades – the faster the wind, the more electricity is produced. The same question also comes up in the offshore versus onshore wind turbine debate. Generally, you get higher wind speeds far out into the ocean, so that would be the ideal place for turbines. However, with increasing distance from the coast, you also get an increased cost – installation and maintenance are much trickier when you need a specialist boat to do them. Onshore wind turbines are cheaper to make and maintain, but they produce less power, and suitable locations for them are harder to

---

[*] The hardest thing about writing *SATC* has been deciding what to leave out. In this chapter, more than most, this proved to be very challenging indeed. Please forgive me if I skipped something that you were hoping to learn about!

[†] This doesn't go on forever. Earth's atmosphere is like a layer cake. The wind speed increase that I'm talking about here happens in the very lowest bit of it – the planetary boundary layer that hugs the surface. Beyond that, the wind speed varies layer by layer.

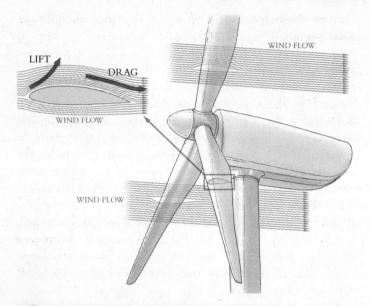

Figure 2.2 Wind turbine blades cut through the air like aeroplane wings. The air on top of the blade is lower pressure than the air under the blade.

find. They also tend to attract a lot of negative press, which we'll talk about in a second.

For now, let's assume we have perfect wind conditions, and a tall wind turbine; we're now part of the way there. We still need to capture the wind and then convert its energy into electrical energy. For the first step, we have to consider the shape of the blades. If you were to cut through a wind turbine's blade, you'd see that it has a similar shape to both a bird's wing and that of an aeroplane. That's not a coincidence. This shape is called an aerofoil because it can slice through air very efficiently.

The rounded section of the blade leads the way and breaks the airflow into two. The air that moves across the flattened bottom part can flow relatively freely, but the air that moves over the top of the blade has a slightly longer path to take. This difference in airspeed produces pockets of different air pressures above and below each of the turbine's three blades. The low pressure on top 'sucks' at the blade, while the high pressure on the bottom pushes it, and the result is a spinning wind turbine. The rotation of the blades spins the loop of wire in the generator, meaning that the alternating current produced by wind turbines can be fed directly into the grid.[*]

The length of the blades is important too because the longer they are, the more wind they can harvest. When a wind turbine rotates, its blades draw out a circle, and with some very quick calculations, you can show that every time you double the diameter of this circle, you increase the amount of electricity you produce by four. So, a turbine with 80m (260ft)-long blades will produce quadruple the amount of electrical power than one with blades 40m (130ft) in length. Because of this, turbines are in the midst of a major growth spurt, with some manufacturers now working on blades that stretch to 100m (330ft) each. A mega-turbine using these blades would have a wingspan equivalent to two and a half Airbus A380 aeroplanes.

'Blades used to fatigue [or weaken] because of the bending stresses they experienced when rotating,' said Dr Kirsten Dyer, a materials engineer at the Offshore Renewable Energy Catapult. 'Today's blades are so large that it's their own weight that causes them to fail, so reducing weight is a top priority.' Today's wind turbine blades are made from fibre-reinforced composites. One example is fibreglass, made from bundles of glass fibre that are woven into cloth, in exactly the same way that bundles of cotton thread can be woven into a fabric. This glass cloth is then dipped into a sticky adhesive that hardens, leaving you with a remarkably strong, lightweight material that can be moulded into almost any shape. However, even fibreglass will be too heavy for tomorrow's mega-blades, so the hunt is on for alternatives.

Another question I've been asked is, **how much electricity can we really get from harvesting wind?** Despite what the critics might have you believe, it's rather a lot. In Scotland in 2014, wind turbines provided enough power to the grid to

---

[*] Because this is all done through a gearbox, each time the wind turbine rotates, it's equivalent to ~1,000 spins of the electric generator.

supply 3.96 million homes. That's equivalent to powering the Chinese city of Shenzhen. In July 2015, Denmark's wind farms made headlines by producing 3.77GW of electricity in a single day – 40 per cent higher than the electricity demand for the entire country. So wind power is no slouch. But I'm not here to tell you that wind turbines are (and I quote Mary Poppins here) 'practically perfect in every way'. They definitely have their challenges, not least their weather-dependent nature. They're also often described as 'inefficient', but that's not strictly true – it is not possible for them to convert all wind energy into electricity. If you could stand directly behind a spinning wind turbine, you'd find that it's rather windy back there. This is because the blades don't remove all of the wind's kinetic energy – if they did, the air behind them would be completely still. Calculations show that even in an ideal world, a wind turbine could never capture more than about 59 per cent of the wind's energy.* Today's best turbines aren't far off that, so in fact, they're doing a very good job.

Some people also think wind turbines are ugly, but there's nothing scientific I can say to change their mind on that. However, there is another, more serious criticism that we need to address before departing from the windy coast and heading to calmer climes. Even the most sleekly designed turbines produce noise, and some believe that it can cause health problems to those living nearby. I'm afraid that this is a question that science hasn't yet found a comprehensive answer to. There is an entire area of acoustics research dedicated to studying the sound produced by wind turbines, and there seems little doubt that their sweeping blades produce infrasound. What's far less clear is the link between infrasound and health complaints.

In 2015, a report from the Australian National Health and Medical Research Council found 'no consistent evidence' to suggest that wind turbine noise causes conditions such as changes in blood pressure, heart disease and depression. The review did suggest however that solid evidence is thin on the

---

* This is called the Betz limit, named after its discoverer, physicist Albert Betz.

ground. The worry with investigations on the effect of infrasound on human health is that there is rarely any neutrality. I'm absolutely not saying that those reporting adverse health effects are lying – we just need to understand that there are limits to what this data can tell us. According to Professor Con Doolan from the University of New South Wales, 'The best way is to simultaneously measure infrasound and personal health data to correlate the effect, if it exists. The type of health data is important too – are the symptoms immediate or do they take time to develop? To answer this, we'll need new multi-disciplinary research, linking engineers with medical and health scientists.' With the cost of wind power decreasing rapidly, we're likely to see more wind farms appearing on the edges of towns and cities. Many urban dwellers will find themselves living ever closer to wind turbines, so these questions are only going to become more relevant.

Solar

The sun, which is an average of 149.6 million kilometres (93 million miles) from the Earth, is our most vital source of energy. Earlier we talked about electrical power, but a more scientific way to think about power is the amount of energy produced or used every second.[*] Taking that definition, the raw power found in sunlight is hard to beat. At midday on a cloud-free day, around 1,000W of solar energy shines down on every square metre of land facing the sun. Let's scale that up and say that the entirety of the pitch of Dublin's Lansdowne Road stadium (currently called the Aviva Stadium) faces the sun. In one hour there'd be more energy shining down on the hallowed turf than is found in 825 litres (181 gallons) of oil.[†] But as we mentioned in Chapter 1, there are many reasons that we can't even get close to capturing all of that energy. For one thing,

---

[*] Yes, it's still measured in watts, but if you're a fan of the SI units, and I most definitely am, you can say that power is measured in joules per second (J/s), where a joule is a unit of any type of energy whatsoever.
[†] A barrel of oil contains about 190 litres (or 42 gallons) and can provide (roughly) 1,667kWh of electricity when burned.

sunlight hits us at different angles depending on the time of year and our location on the Earth. It's also not midday all the time, and if it is cloudy, the amount of sunlight that reaches the surface decreases further. Even with these limitations, it makes a lot of sense to tap into our power plant in the sky, and green plants are expert at doing this. Through the process of photosynthesis, they convert solar energy into chemical energy, which allows them to make sugars – their source of food. For cities, we're really interested in transforming sunlight into two things: electricity and heat.

Given that shiny, high-tech solar panels have become ubiquitous on the rooftops of urban areas, we should probably discuss how they work. Systems that convert sunlight into electricity are called photovoltaic cells, and the vast majority of those you see on buildings are made from silicon. It's found in the heart of every electronic device, but it's not as simple as just digging this bluish-grey material from the ground beneath your feet.* The silicon we use in electronics and solar cells needs to be carefully designed to do the job.

Silicon is a semiconductor. Metals like copper are conductors, which means that electrons flow through them very easily. Insulators, like rubber, don't allow electrons to flow at all. Semiconductors lie somewhere in the middle, and with some clever material science, they can control how electricity passes through them. In solar cells, we tend to use a sandwich of silicon:

1. The bottom is a layer of silicon that is missing a few electrons – this is called positive, or **p-type**.
2. The filling is a layer of silicon that has too many electrons – this is called negative, or **n-type**.
3. The top is a layer of anti-reflective glass.†

---

* By weight, silicon is the second most abundant element in the Earth's crust, with oxygen taking the top spot.
† Reflections are the enemy of solar panels; any light that is being reflected is not being harvested by the device, and therefore not being converted into electricity.

Sunlight is made up of tiny particles called photons, and when they hit the solar cell, they knock out the electrons in the n-type layer.[*] Because of the gaps in the p-type layer, these electrons flow, producing current electricity. By putting lots of these solar cells together, you get a panel that can produce electricity from sunlight. There is one other thing to consider: sunlight isn't all one colour, as we know from rainbows. It's made from light at different wavelengths. Only certain wavelengths can knock out electrons and cause them to flow, and this limits how much of the sun's light can be turned into electricity. But, as we'll discover in just a few pages, there's some awesome research looking into how we can change that. In reality, today's very best commercial systems can only convert about 20 per cent of the sun's energy into electricity.

Bearing all that in mind, how much electricity can we expect to produce from photovoltaic cells? I played around with an online calculator (called PV Watts) from the National Renewable Energy Lab and it shows that location is everything – the same solar panel in Los Angeles will produce almost double the amount of electricity (in kWh) than if it were in Oslo. Because we want photons to smack into electrons for as much of the day as possible, the panel's angle to the sun is also important, and we'll need a way to store that electrical energy, so that we can tap into it when the sun goes down. While they might be getting much cheaper, today's silicon solar cells are still expensive, and remain unaffordable for many city-dwellers. So, it's fair to say that Elon Musk glossed over a few points when he announced (in 2015) that the US could easily get all of its electricity from photovoltaic solar panels.

Of course, electricity isn't the only option for solar energy – sunlight can also be converted into heat. **Solar thermal** systems have been used for decades, and while they look a bit

---

[*] Light can be described as either a wave or a particle, so to make my life easier, I'll switch between these descriptions throughout. Purists, forgive me!

like standard solar panels, they're quite different. They make use of the fact that as well as providing light, the sun also gives us warmth. Brazilian cities have heavily invested in solar thermal power. On the tops of tall buildings in São Paulo and Rio de Janeiro, hundreds of dark glass panels can be found. They are called collectors, and as the name might suggest, they collect the heat energy that comes from the sun. This heat gradually warms water that is being pumped through pipes under the panel, which in turn is used to heat up the water stored in domestic boilers, et voilà! Hot water, ready to be used. Systems like this can provide up to half of the hot water needed in a typical urban household, so they are extremely useful.

If your building is a bit larger or you need higher temperatures than that of your average shower, solar concentrators can help. By focusing the light into a tight beam, they can make cells more efficient, but they also produce a lot of heat (if you've ever started a fire using a magnifying glass, you'll be familiar with this effect). Using a specific piece of glass called a Fresnel lens (more on this in Chapter 4), rooftop systems can comfortably heat water beyond its boiling point. This makes them very popular for use on apartment and office blocks. Research funded by the Brazilian government found that in areas of high sunshine, these concentrator systems could save upwards of 78,000kWh a year, equivalent to the amount of energy found in 8,900 litres (1,960 gallons) of oil.

Cities have a huge role to play in the future of solar power, and in fact, they could become the dominant source of solar energy – if deployed correctly. A report published by the Environment America Research & Policy Centre showed that in 2014, just 20 cities in the US accounted for 6.5 per cent of the country's solar capacity, using just 0.1 per cent of its land. A study from Stanford University estimated that investing in solar power (both photovoltaic and thermal) solely in the cities of California could provide the state with all its electricity and hot water needs, at least three times over. And yes, these figures do take into account the intermittent nature of solar power, and the performance of current systems. Even with those limitations, it still makes sense for many.

As you've probably gathered by now, the system that exists beyond your plug socket is very complex indeed... especially when we consider that we have only discussed three ways to produce electricity. The key thing to remember about the grid is that supply must always, always match demand, and this has huge implications for intermittent electricity sources such as solar. But as we now glance into the future, we'll see that tomorrow's grid will be very different from what we have today.

Tomorrow

There are three things we know for sure about the future 'electric city':

1. cities are the dominant force behind economic growth,
2. most are looking towards renewable energy, and
3. as our populations grow, so will our hunger for electricity.

Given all that, **what can we do to provide what growing cities need, while lessening our impact on the environment?** There is some truly ground-breaking research being carried out to answer that very big question. These are new technologies and approaches that will, genuinely, change the world. In the process of reading countless papers and reports to understand the current situation for urban energy, I was lucky enough to speak to some of those leading the way. Given that there is nothing I love more than talking to very clever people, I was in my element. Let's wire up our city, and start by looking from the top down – are there ways we can design our cities so that it's easier to manage our energy supply, demand and footprint?

Design

Yes, OK, that was a leading question. There's a lot of evidence to suggest that the physical layout of cities has a huge impact on their energy usage and greenhouse gas emissions. In early 2015, a huge study was published by a group of German and US environmental scientists. They looked at 274 cities, including all of the world's megacities (those with more than 10 million people) and found that 'if current trends in urban expansion continue, urban energy use will increase more than threefold... by 2050'. Worrying, I know. The good news is that by analysing different types of cities, it's possible to identify ways to mitigate this impact. Those cities that are compact, with lots of transport options and many mixed-use buildings, were found to be considerably 'greener' than cities that sprawl. A recent study from the University of Georgia also found that cities with a large number of green spaces tended to be cooler than those without. This is all part of the so-called urban heat island effect that we'll come back to in Chapter 3. The basic premise is that because we use dark, heat-absorbing materials in our cities, they trap a lot of heat. If instead, this heat energy was absorbed by trees and vegetation, it would evaporate water from them, creating a natural cooling effect. More trees please!

In terms of powering tomorrow's cities, few of the best options will be cheap to implement, so you might wonder if it's worth it. Research from the University of Leeds looked at the direct costs of low-carbon investments for cities, as well as the financial returns. It turns out that these technologies could pay cities back *17 times over* (for every $1 invested, cities would get a return of $17). This study has made some assumptions around future energy prices, and on how quickly the price of low-carbon technologies will fall, but the method used looks pretty solid, and some have suggested that it may even underestimate the return on investment. Speaking to *BusinessGreen e-magazine*, the report's lead author, Professor Andy Gouldson said, 'Cities around the world can commit to really ambitious carbon reduction targets, safe in the knowledge that economically, they will more than pay back.' So, time to dive into the technologies themselves.

Hub

I spoke to a lot of people while researching this chapter, and there was one thing that everyone agreed on: today's electricity grid is a dinosaur. To design a system that links millions of households to a reliable source of electricity is a remarkable engineering (and logistical) achievement, but it simply won't be fit for purpose in the cities of the future. As demand grows, and the sources of electricity become diversified, we will need a suitably futuristic system.

Let's start with generation. Something that's rarely spoken about is that the grid's many steam-powered, endlessly spinning generators do more than simply spit out a stream of electricity. They actually offer a physical stability to the system. It's a bit like a spinning top – once you get it moving, it keeps on spinning. With something the size of a national grid, you might have hundreds or thousands of spinning turbines connected to each other. Their rotational energy means that even if a plant goes offline, its generators keep producing electricity for a while. This gives the grid managers time to redirect power from elsewhere, stopping a potential blackout. Balancing a grid is extremely challenging, and balancing one that includes variable sources such as solar and wind power is more challenging still. A widespread adoption of sensors across the network, known as the smart grid, will be vital to managing that challenge. Not only will these be able to monitor varying supply and demand, but they could also automatically adjust those variables, to keep everything balanced all the time.

Longer-term, we'll see closer links between generation and distribution. Called (unsurprisingly) distributed generation, this is likely to revolutionise the way future cities interact with the grid. Instead of simply tapping into it, we'll see more of a mutual exchange. 'The future grid doesn't look anything like today's – many more households will have generation or storage systems on-site, and this will change the game,' said

Professor Win Rampen from the University of Edinburgh. These 'mini-grids' would not only be able to integrate with the main grid, but they could collect, manage and distribute locally generated electricity to surrounding buildings. This decentralisation of electricity is already happening, as we touched on briefly in Chapter 1 – combined heat, power and cooling (CHPC) systems are beginning to find their way into cities.

CHPCs are effectively small power plants, which instead of wasting heat energy, use it to provide hot water to the building. Because they're producing both electricity and hot water, more of the 'energy in' is being transformed into useful 'energy out', making CHPCs substantially more efficient than conventional power plants. Depending on the season, Shanghai's Pudong Airport gets around a quarter of its electricity and up to half of its heating from its on-site plant, and when applied to a whole district, the benefits can be enormous. According to UNEP, 'In Gothenburg, Sweden, district heating production doubled between 1973 and 2010, while CO_2 emissions fell by half, and the city's nitrogen oxide (NO_x) and sulphur dioxide (SO_2) emissions declined even more sharply.'

The growth in electric vehicles on our roads will be another key component to future distributed generation – after all, they are effectively moving batteries that need to be recharged. Senan McGrath is a leading figure in the development of a Europe-wide infrastructure for electric vehicles. When I chatted to him about this, he made the challenge clear: 'If every car in Europe suddenly became electric overnight, the grid would have to provide 20–25 per cent more electricity than it does today to power the fleet.' McGrath also pointed out that with a clever use of sensors and a subtle change of habit, it would be more than doable: 'We've calculated that with smart-charging – incentivising the people of Europe to charge their cars off-peak – we could manage this without building a single plant or raising a kilometre of transmission lines.'

It's clear that our relationship with electricity will change, in terms of both how we produce it and how we use it. It will definitely become more complicated before it gets simpler (much like life), but I think that those cities already investing in the 'smart grid' approach will reap the benefits, sooner rather than later.

Blow

Wind power is undoubtedly going to play a big role in this future grid, especially in those cities with decent wind speeds or a coastline. But if you've ever shaken your head in disappointment at the sight of a stationary wind turbine on a moderately breezy day, you'll know that they don't always work. There are many reasons for this – a mechanical fault in the gearbox, or too much or too little wind, or it's been turned off because the grid is over capacity. A Scottish company called Artemis may have found a way to solve many of these issues.

In their system, the rotating turbine blades don't drive a gearbox. Instead, they drive a very large 'ring-cam' pump, so called because it is made from a circular array of pistons. As the turbine turns, each of these pistons moves in and out, pumping oil around the system and causing the generator to spin. Hydraulic systems like this have been used in industry for decades because they're lighter than a gearbox, but they've never been very efficient. By adding some fancy control electronics and software, different pistons in the pump can be turned on and off as needed, making it more efficient. It also means it's more suited to variable wind conditions. Jamie Taylor from Artemis told me, 'Say the turbine is only turning at half of its maximum speed because it's not very windy. This system can still capture most of that turning force, whereas traditional systems might only manage a small fraction.' Artemis's ring-cam is being used in a 7MW turbine just off the coast of Scotland, and its performance has been attracting attention.* It's not an exaggeration to say that this

---

* Let's take the UK government's latest figures on 'capacity factor' (*i.e.* how much electricity a wind turbine produces) at 30.2 per cent, and London's average annual electricity consumption at 10,900kWh. In that case, a 7MW turbine could provide enough electricity for around 1,700 households.

seemingly unsexy technology could change the face of wind
power. Which, to me, makes it very sexy indeed.

There are countless new technologies that claim to harvest
wind in various ways, from building facades covered in
flapping bits of plastic, to vibrating, bladeless structures. I'm
not saying they don't work – in fact, there's scientific merit in
both of these ideas – but we should always don our sceptic's
hat when reading stories like these. The bladeless system in
particular made a lot of headlines, so let's unpick it. First,
they look a bit like a scaled-up version of the 2012 Olympic
torch. The idea is that instead of capturing clean, fast wind
(like standard turbines), these systems use turbulent wind to
vibrate the whole structure. The movement is then
transformed into electricity using a non-rotating generator.
So far, so plausible. But one of the company's big selling
points is that their structures do all this in complete silence
– and there go the alarm bells! Vibrating things generally
make noise, so it's not reasonable to say otherwise. Speaking
about this technology, Professor Sheila Widnall from MIT's
AeroAstro division said, 'The oscillating frequencies that
shake the cylinder will … sound like a freight train coming
through your wind farm.'

Another 'green' technology that recently caught my eye
was turbine trees. These harvest wind using miniaturised
versions of the spinning signs you often see outside shops.
Officially called vertical-axis wind turbines, 72 of these
aeroleaves have been added to a steel 'tree' to produce a wind
power system that, I must admit, looks beautiful. The inventor,
Jérôme Michaud-Larivière, designed them specifically for use
in urban environments, but they only produce 3.1kW – maybe
enough to power three or four streetlamps. Realistically,
they're unlikely to make a dent in the electricity footprint of
your average city.

Because of its variability, wind-generated energy can be
tricky for electricity grid managers, so most set a maximum
limit on how much of the grid can be powered by it at any
given instant. In Ireland, that limit is around 50 per cent, and
Senan McGrath told me that, 'Right now, our transmission

operator forces wind generators to switch off once we go beyond that. But we're all committed to increasing this proportion to 75 per cent in the coming decade.' In the meantime, there is one very clever way to harvest this 'wasted' wind energy, without unbalancing the electricity grid. The German city of Mainz uses it to produce hydrogen.

Hydrogen is the lightest, simplest element we have, but because it likes to bond to other stuff, we have to make a special effort to extract it. Water is the ideal compound for this – as we'll talk about in Chapter 3, every water molecule in the universe is made from two hydrogen atoms and one oxygen atom (hence, H_2O). In the Mainz plant, huge systems called electrolysers use (you guessed it) electricity to break the water molecule apart, sending negatively charged oxygens in one direction, and positively charged hydrogens in the other. The hydrogens then join together to form hydrogen gas (H_2), which can be collected.* As we'll discuss in Chapter 5, hydrogen gas will soon have a much larger role than it has today – it may just power the car of the future. Generally, electrolysers require a lot of electricity to produce hydrogen, which would make hydrogen fuel cell cars just as bad as any others. But by tapping into wind power that would otherwise be wasted, cities like Mainz may have solved at least part of the challenge of a changing, turbulent grid system.

Shine

When I asked Professor Win Rampen about the future of the urban grid, he said, 'It's always difficult to predict this stuff, but inevitably, we will need to learn to live off the energy we get from the sun.' As of 2014, the landscape for solar power looked a bit samey, with silicon making up 92 per cent of global solar cell production. You might expect that this is because it's the best material for the job. In reality, it's because in the 60-plus years since scientists invented the first silicon transistor (effectively a tiny switch, which is the foundation of

---

* The oxygen gas can be collected and used in numerous industrial applications too.

modern-day electronics), we've become very good at making things with it.

However, one thing that silicon's not particularly good at is converting sunlight into electricity. Like all semiconductors, it performs best at a single wavelength of light, but unfortunately, silicon only likes long wavelength, low-energy light.* Given the fact that we can only convert the energy we absorb, this isn't ideal. Don't get me wrong, silicon has its advantages – we can produce it *relatively* cheaply, it's *relatively* efficient and it's *relatively* abundant, but it could never be described as the ideal material for solar cells. Dr Chris Case, the Chief Technology Officer of Oxford Photovoltaics (Oxford PV) summed it up nicely when he told me, 'Silicon was a well-understood, readily available option, rather than it being the best possible option.'

So if we're going to populate our future cities with solar cells, what are our options in terms of materials? For solar thermal (sunlight into heat) applications, the future looks mostly nano. A nanomaterial is any material in which at least half of its particles measure between 1 and 100 nanometres (nm) in length. To give that catchy definition some context, your standard 30cm (12in) ruler is 300,000,000nm long, so we're talking about some very small particles here. Researchers at the University of California, San Diego have designed nanomaterials that are particularly good at harvesting the sun's heat energy. If you've ever worn a black T-shirt on a sunny day, you'll know that dark colours warm up very quickly. The warmth you feel is actually a wavelength of light that we can't see, called infra-red, which dark colours are particularly good at absorbing.† To make use of the effect, in San Diego Professor Sungho Jin created several dark, textured coatings made from tiny particles of different semiconductors, to produce 'a material that absorbs sunlight [and] doesn't let

---

* Silicon tends to absorb red light, which has a wavelength of 620–750nm.
† In fact, black materials are good at absorbing light at all wavelengths – that's why they look black!

any of it escape'. These coatings have been shown to absorb 80–90 per cent of the sunlight that hits them, while reflecting very little – perfect for solar thermal systems. Although much work is still needed to transform these materials into real, commercial products, the research is well on its way. Existing solar thermal systems are already pretty good, but these black traps for sunlight could improve their performance considerably.

Nanomaterials are also making their mark on solar photovoltaic (sunlight into electricity) applications, but for a different reason. In conventional solar cells, to make one electron move, you need to hit it with one photon of the right wavelength. So lots of sun = lots of electricity. But when you produce tiny, nano-sized crystals of semiconductor materials, something weird happens. Instead of **one photon, one electron**, you get **one photon, multiple electrons**. The reason for this isn't yet fully understood by quantum physicists, but it's an effect that's already being used to create more efficient solar cells. The other bonus is that by playing around with the crystal size, you can make solar cells better at absorbing multiple wavelengths of light. We could also go ultra-thin – materials made from thin films of semiconductors (even 'ye olde worlde' silicon) are being investigated in research labs across the world. As well as their potential to provide flexible solar cells, these materials can, in theory at least, absorb light much more efficiently than silicon. Right now, commercial thin-film solar cells still underperform, but that's more to do with the challenge of manufacturing them, rather than the materials chemistry itself.

So, could we ever clad entire skyscrapers with ultra-efficient solar panels? Perhaps surprisingly, the answer is 'maybe', and it's thanks to a type of material that combines nanocrystals with thin films. Perovskites are the latest buzzword in solar power. Named after a Russian mineralogist called Lev Perovski, perovskites might be made from any number of materials; it's the crystal shape that they all have in common. Originally suggested for use in solar cells in

2009, their development has skyrocketed since. In just six years, their conversion efficiency (that is, a measure of how good they are at turning sunlight into electricity) increased from below 5 per cent to over 20 per cent, leaving them knocking on the door of the best comparable silicon cell. When combined with silicon in a structure known as a tandem solar cell, they could become the most efficient solar photovoltaics in the world. Even a layer of perovskite just 300nm thick can absorb sunlight, which means they can be semi-transparent too.

Oxford PV's Dr Chris Case told me that they've carried out some initial modelling on the potential use of these solar cells in building windows. 'This technology could eventually turn skyscrapers into vertical solar power plants. Even taking into account real measures of UK sunlight and shadows cast by surrounding skyscrapers, London buildings such as the Cheese Grater could produce 1GWh [1,000,000kWh] of electricity.* For a typical skyscraper, this could be up to half of its electricity demand. In terms of when we're going to see this, it's definitely a way off (maybe 15+ years). But when it does happen, it's going to be very cool.

Of course, solar panels and wind turbines have something in common: neither source produces electricity constantly. Wind turbines need moving air and solar panels need sun, and this has led to huge criticisms about investing in them. My take on it is that no single solution exists to solve the energy crisis; anyone who tells you otherwise is lying. Renewable energy sources like wind, tidal, solar and geothermal are much more sustainable than those based on fuels mined from the Earth and then burned. And as we discovered just a few pages ago, even the money involved shows that renewable technologies are a no-brainer for cities. But we already know that today's grid isn't set up to have lots of these variable sources plugged into it. So for much of our future electricity needs, the biggest challenge will be to find cheap, plentiful ways to store this energy.

---

* Officially known as the Leadenhall Building.

Store

As it stands, much of the electricity generated within the grid must be used immediately, because we don't have anywhere to put it. This seems very odd even today, thanks to the always-on-the-go nature of city living, and in tomorrow's cities it will seem downright archaic. Creating high-capacity energy storage options is not as simple as it seems (as smartphone owners will know all too well), but there are lots of potentially exciting options.

First, we don't just need to store energy in the form of electricity. For solar farms that concentrate the sun's light into focused beams, it's all about heat energy. For this, you need materials that can cope with incredibly high temperatures. Historically, we've used the rather sci-fi sounding molten salts. If you could take common table salt and heat it to a toasty 801°C (1,474°F), you'd be left with a liquid that looks a bit like water. But, unlike water, it doesn't turn to steam, and this is the basis of thermal storage – the heat energy in the sunlight warms up huge, insulated vats of salt, which store that heat for many hours. In the meantime, the heat is used to boil water, which is then pumped to the city for us to use. Another similar option is to use phase change materials, which can store and release heat as they change from solid to liquid and vice versa. Because they do this at lower temperatures, they are perfect for standard solar farms. We can also capture the energy hidden in moving water. Hydroelectric power involves pumping water uphill before letting it flow downhill, where it then spins a generator and produces electricity. Believe it or not, as of 2012, hydroelectric accounted for 95 per cent of the world's bulk energy storage capacity, and it's poised to remain just as popular in the future.

Of course we can store electricity as chemical energy too, in batteries. Elon Musk made (more) headlines in 2015 when he launched the Powerwall – a battery for your home, wired up to your rooftop solar cells. When the sun is up, the panels charge the battery, and during the night, the battery powers your home. This is a very good idea, but it has some flaws. London houses use, on average 1.24kW every hour.[*] A fully charged Powerwall (costing \$3,500 (£2,500) at the time of writing) can provide 1kW on average for 10 hours, so let's be generous and say that the capacity is about right. But to get you entirely off the grid (and that seems to be the intention), you'd need enough solar panels to both power your home during the day *and* charge the battery, so that it can run all night. That's a lot of solar panels! As well as that, batteries produce direct current (DC), whereas our gadgets run on alternating current (AC) – we'd need to factor in the cost of a converter, which runs to thousands of dollars. I'm not saying don't bother, but with all things considered, I think these systems will remain a toy for environmentally friendly millionaires, at least for the next 5–10 years.

One of the challenges with batteries, no matter what their size, is that their lifetime depends on how often they go from completely empty to completely full. This is called a cycle. If you charge your smartphone every day, you probably have about 18 months (or 500 cycles) of your battery performing at its best. After that, its capacity (the amount of charge it can hold) gradually begins to decrease. There are a few materials under investigation that might change that. One of them is probably the most-researched material today: graphene. Discovered in 2004 by two Manchester-based scientists, Professors Andre Geim and Kostya Novoselov, graphene is a single layer of carbon atoms arranged in a honeycomb pattern.[†] To be honest, it looks a bit like tiny chicken-wire. Graphene is the thinnest material on Earth and despite being only one atom thick, its tensile strength is higher than that of steel. It is also incredibly conductive, which makes it very interesting for battery researchers.

---

[*] This is very much an average – there are peaks and troughs in electricity demand, but the latest figures say that the average home uses 10,900kWh per year, so I worked from there.

[†] Andre Geim is a bit of a legend. He once named his hamster as a co-author on a paper without anyone noticing. He also demonstrated a super-sticky coating by printing it onto the hands of a Spiderman toy, and used magnets to levitate a frog.

One of my favourite recent developments in this area includes a battery that combines graphene with aluminium foil. It is super-fast to recharge and it can cope with more than 7,500 cycles before experiencing any lasting damage. There is a catch, though – the capacity of the foil battery was considerably smaller than that of a smartphone, so this won't be a practical option for large-scale energy storage. But it's not just academics who are investigating graphene. In 2015, researchers from Samsung replaced the electrodes in a standard lithium-ion battery with ones made from graphene-coated silicon. Because the electrode is the conductive rod that allows electric charge to enter or leave the battery, this change in design almost doubled the battery's capacity – music to the ears of those worried about our electric future. All this is at the research stage for now, but it is an area that is growing rapidly, so don't be surprised to see real-world graphene batteries within the next decade.

There are lots of other ways that we will store energy tomorrow, but I think I have thrown enough at you for this chapter. The fact is that we're right on the edge of transforming the grid for ever. As we move to the city of the future, we will depend far less on a centralised system of electricity generation, and look to finding clever ways to produce and distribute electricity locally. Few cities will be able to entirely 'cut the cord' from the grid, but most will become mutual partners in it, making electricity a matter of give and take. All of this will be enabled by the many new storage technologies currently in development, combined with vastly improved and more sustainable systems for generating our electricity.

We've come pretty far in this chapter, from meeting the electron and Nikola Tesla, to designing energy-efficient cities and skyscrapers-cum-solar farms. But in our journey around the city, we've only just begun. Next is something so vital to life that we search for it on other planets. Time to explore the pipes under the city streets.

CHAPTER THREE

Wet

Water is everywhere. It covers almost three-quarters of Earth's surface, and makes up two-thirds of our bodies. This remarkable chemical has helped to shape life on our home planet. Culturally, water is hugely significant too, with countless civilisations built on or near waterways, and many vicious battles fought to control it. Today we're fighting a different battle, as unparalleled population growth and a changing climate have led to water shortages across the globe. Yet we continue to flush huge volumes of the very cleanest water down the drain.

That's not all we have to worry about either. We are famously becoming a 'disposable society', generating more waste than humans have ever done before, contaminating our water, and adding to landfills already the size of small towns. You might be thinking that we are doing a pretty shoddy job of managing our water and waste, and you'd be half right. While it is absolutely true on a global scale (and I'm going to quote some pretty scary figures at you in a minute), locally, things *are* changing – individual cities are using science, engineering and technology to protect their precious water supplies, and to get much more from what we throw away.

Today

Turn on your tap. Now, flush your toilet. If you can do both, you should consider yourself very fortunate. According to the World Health Organisation, more than 630 million people don't have access to clean water - that's almost twice the number of people living in the United States. In 2014, the UN announced that a staggering 2.5 billion (almost a third of the world's population) live without adequate sanitation. If you are one of the lucky two-thirds, you may get through anywhere between 130 and 300 litres (29 and 66 gallons) of water every single day. We drink it and bathe in it, we use it to heat our homes, manufacture products, and to grow and cook our food. Extend that to the population of your nearest city, and you begin to appreciate how big the task is for urban water suppliers.

According to the World Bank, our cities also generate about 1.3 billion tonnes of solid waste per year. Unsustainable, by any measure. There's also a lot of different types of waste to consider – from food and faeces to mobile phones and wet wipes. All of them come with different challenges and different scientific solutions. We're going to talk about how individual cities manage all this, following water from the source to its first stop in the urban home. We'll then take a (perhaps less pleasant) journey at the other end, to discover where all our waste goes, and to learn how tomorrow's technologies could change everything. As always, we'll meet lots of experts along the way, to help us understand what really goes on in the pipes under our feet.

But first, a brief historic interlude. Although they weren't the first to turn their hands to plumbing, the Romans definitely made a name for themselves in the 'spectacular water-related infrastructure' sector – from aqueducts over inconveniently placed valleys, to underground pipes that directed water and waste to where it was needed. We can also thank them for the words 'plumber' and 'plumbing', which both stem from *plumbus*, the Latin word for the metal, lead.* This isn't a coincidence, of course. It was this seemingly magical ingredient that put the Romans miles ahead of anyone else, for over a thousand years.

Pipe

Lead is a rather interesting element. First, its atoms are huge, each one with 82 electrons spinning around a nucleus of 82 tightly packed protons and 126 neutrons. This large mass

---

* This is also why the chemical symbol for lead is Pb. You can thank me for that in your next pub quiz.

means that it's really good at stopping things like x-rays. It is also soft, easy to melt and can be moulded into various shapes. Compounds of it have helped us to create the materials that heralded the Information Age and allowed us to develop the first practical batteries; but for our ancient ancestors, lead was a wonder material.* At various points in time, it was used to sweeten wine, made into cosmetics, and was a key ingredient in cooking vessels. It was also because of lead that the world got its first robust water system. It was the ease of shaping it, joining it and repairing it that made lead the material of choice, alongside clay, for Roman water and waste pipes.

Fast-forwarding to London in the sixteenth century, it was wooden pipes that were all the range – a trend that extended to Boston and other US cities in the 1800s. At around the same time, plumbing began to enter the (second) Iron Age. Cast iron started to supplant earlier material choices, and many of today's greatest cities still have those pipes to thank for their growth. But what prompted the move away from lead? Short answer: considerable health concerns. Useful as it is, lead is highly poisonous when swallowed or inhaled. Long-term exposure to it has been linked to conditions that affect the nervous system, the cardiovascular system and the immune system. Despite this, it remained in use for much, much longer than you might expect – in fact, lead pipes are still found in most cities, among the more modern options of copper, concrete and plastic.

Before we delve into today's pipe materials, we need to talk about leaks. No matter how modern it is, no city is immune to them. In London, official figures say that about a quarter of the city's potable (drinkable) water is lost through leaky pipes. In Johannesburg, it's about a third. New York's soon-to-be-replaced Delaware Aqueduct loses somewhere between 3 and 6 per cent every day due to leaks.†

---

* It's never been used in pencils – the soft grey material is (and always has been) graphite.
† The Delaware Aqueduct supplies half of New York's water – that's 2.27 billion litres (500 million gallons) every day.

That 'lost' water could fill 150 Olympic-sized swimming pools; it's not as insignificant as the percentages might suggest. Leaking pipes are a huge problem in our cities, but how can technology help?

The first, and most obvious, option is to replace the pipes with more robust ones. London's cast iron Victorian water mains have a bit of a bad reputation, constantly bemoaned by residents and urban planners alike. It's all due to the cast iron pipe-killer, corrosion; where strong, shiny metals gradually turn into weak, crumbly metal oxides by reacting with water and oxygen. These molecules work as a team, ripping electrons away from the metal atoms and forming other compounds. In iron, what you're left with is rust, and though it may not seem like it, this is iron in its more stable form, iron oxide.* Corrosion is even more pronounced if the water is acidic, or contains salt, because it speeds up the electron-ripping-and-bonding cycle. It is possible to slow down corrosion in pipes by coating or painting them (to be discussed more in Chapter 4), but it's always there, chemically eating away at the metal, slowly making it weaker. For the Victorian pipes of London, this chemistry has been going on for 150 years, so it's not all that surprising that they're failing today.

There is no 'one pipe suits all' for water and waste. Today, cities use a combination of different materials in different parts of their plumbing systems. Compared to other metals, copper is relatively corrosion resistant, making it suitable for use in watery environments. Oh, and here is a 'did you know?' moment: copper is antimicrobial. A paper published in 2011 concluded that 'Bacteria, yeasts, and viruses are rapidly killed on metallic copper surfaces', and this result has been repeatedly backed up by research from the US Environmental Protection Agency, among others. While the exact mechanism behind its antimicrobial behaviour is under debate, it seems as if the copper can penetrate a cell's protective

* Iron likes to bond to things, so to use it, we must first physically extract it from its ore, haematite (Fe_2O_3); when it rusts, it's really turning into a compound similar to its original form.

membrane and damage its DNA. The flip side is that in very high concentrations, copper can be damaging to larger organisms, such as trees or humans, so we need to carefully monitor its use in water systems.

Then we have plastics. There are so many different plastics used in municipal water systems that I could write an entire chapter just on them. Instead, here a few titbits of information. First, plastic isn't quite the right word – instead we should say polymer. This word literally means 'many parts' and it refers to any material made from very long chains of repeating molecules. Polymers can be found in everything from bouncy balls and glue to lunchboxes and plumbing. Their properties depend on the type of molecules in the chain and the way the polymer is processed. But something you may not realise is that the vast majority of these polymers are extracted during the production of oil and other fossil fuels, making them rather unsustainable.

Things are beginning to change on that front (and more on this when we glance into the future), but for now, we're stuck with them. Most pipes are made from polymers called thermoplastics – they soften when they're heated above their melting temperature, and can be easily shaped when squishy. Because of this, they can be recycled – melted, moulded and hardened a large number of times. Thermoplastics combine high strength and flexibility in a lightweight, cheap material, and they're resistant to a wide range of chemical reactions – useful when you consider the diverse liquids they need to carry in our water and waste systems.

This brings us nicely on to our next topic: water itself. It's time to start our first journey, following a few lowly water molecules all the way from source to tap. Have you ever wondered where your city gets its water from? Maybe you think you already know, but I'm here to tell you, you're probably wrong.

H_2O

Made up from only three atoms, water is a seemingly simple molecule, but it's actually a bit weird. If it behaved like other molecules of the same size, it should be a gas, rather than a liquid, at room temperature.* So why isn't it? The scientific explanation is hydrogen bonding, but really, it's about that feeling of being irresistably drawn to one another.

When water's three atoms come together, they bond into a roughly pyramidal shape, by sharing their negatively charged electrons. But it's not an equal share – oxygen (at the top of the pyramid) holds the electrons a bit closer, which makes that end of the molecule slightly negative. It leaves the hydrogen end (the base of the pyramid) slightly positive. Throw lots of these polar molecules into a cup, and they act like tiny magnets – the negative pole of one is attracted to the positive pole of another. This is the process of hydrogen bonding and it means that water molecules are incredibly attracted to each other, even at relatively high temperatures. It's because of these bonds that we have liquid lakes and oceans on Earth, so they're kind of a big deal. Another weird thing about water? It expands, rather than shrinks, when it freezes – this is why water pipes can sometimes crack in winter.†

Anyway, now that we're a bit more familiar with the chemical itself, let's put it into the context of a city. Ancient civilisations thrived or perished by how easily they could access water. With limited options for transporting it, water needed to be close by, reliable and clean in order for a community to survive. These days, our cities can source their water from further afield, but we are still, without exception, at the mercy of our water supply. Some cities have gone to rather extreme measures to manage the balance between water and waste.

It's the mid-1800s. Chicago's population is booming and the water system is starting to struggle. As was common at the time, the city's sewers discharged their waste directly into the river, which flowed down to Lake Michigan. The problem was that the lake also acted as the city's drinking water supply,

---

* Room temperature is loosely defined as around 20°C (68°F).
† To read more about how this simple fact helped life to flourish on Earth, I'd recommend reading *The Water Book* by Alok Jha.

and when the sewage and drinking water mixed due to poor separation between them, the water became contaminated. Waterborne diseases were estimated to have killed 90,000 Chicago residents in a 50-year period, so the city came up with an ambitious plan to sort it out. They would reverse **the** flow of the river, and they built a huge canal to do the job. It used a series of locks – effectively, water elevators – to keep the lake higher than the Chicago River, forcing the river to flow away from the lake rather than towards it. Together with two other canals and sewage treatment plants, Chicago vastly improved the water quality for the city. A huge engineering achievement? No doubt! But in reality, it didn't solve the problem, it just moved it further downstream to the Mississippi, causing flooding and probably contaminating other water supplies. In addition, these new routes helped invasive fish species to enter US waterways, threatening the ecosystems of the Great Lakes. A vast, expensive construction project to correct these issues is only now in the works, but it could take up to 25 years longer to complete.

This all goes to show that the management of water and waste is incredibly complicated – according to Sandy Lawson, senior water engineer at Bechtel, 'it's really not a case of simply finding some freshwater and pumping it into our cities'. All water on Earth forms part of a vast water cycle, so if we're going to talk about supply, there isn't an obvious starting point. Molecules of water are continuously moving, changing between ice, liquid and vapour. Water may be temporarily stored in humans, or in plastic bottles on supermarket shelves, or equally, it may fall as rain, soaking you as you walk home.* The timelines for each of these stages vary hugely – a water molecule might spend 10 days in the atmosphere before falling as precipitation, or it could last a

---

* According to the US Geological Survey, all of the water – solid, liquid, gas – 'on, in and above Earth' would fill a space 1.39 billion cubic kilometres in size – an incomprehensibly large volume!

month or so in soil, or spend thousands of years in the ocean. It is a very dynamic system, but the total volume of water we have available is essentially fixed. And, perhaps surprisingly, only 2.5 per cent of the total water budget is freshwater – the salt-free version that we need to survive.

The obvious sources of freshwater are our rivers, lakes and man-made reservoirs, and most historic cities still rely heavily on these; for example, the Nile provides almost all of Cairo's water. But there is far, far more freshwater under our feet than there is on the surface. Aquifers are vast areas of rock (usually sandstone or gravel) that naturally store so-called groundwater. Because they're full of tiny holes, aquifers allow water from rain and snow to trickle through and fill up the gaps. This natural 'filtering' process also means that groundwater is pretty clean and needs minimal treatment before being safe to drink. To access it, wells are drilled into the aquifer, and the water is pumped out. Most major cities in the US depend heavily on groundwater, and it's also the source of much of the world's bottled water. But it is possible to suck aquifers dry if they are not carefully managed. In cities, this is a particular problem because we tend to use lots of impermeable materials, like tarmac and concrete. If water can't get through the surface, it can't reach the aquifer to replenish it. Add to this the changing weather patterns we see, and many cities are using groundwater faster than it can be replaced.

Not all aquifers are buried deep underground. Miami's Biscayne Aquifer is closer to the surface than most, and provides a large proportion of the city's drinkable water. But there are widespread concerns that it is losing a battle with salt. When areas of marshland around the city were drained a hundred years ago, seawater began to intrude into the gaps. Control structures were built to stop it, but as sea levels have continued to rise since, more salty water is finding its way into the aquifer. Despite plenty of rain, the aquifer's water may eventually become undrinkable. In cities across the world, contaminated and dehydrated aquifers are a growing challenge, but, as we'll discover later, there may be a way to redress the balance.

Perhaps surprisingly, the vast majority (around two-thirds) of the world's freshwater is locked up in snow, ice and glaciers. They act like frozen reservoirs, gradually building up during the winter before melting throughout spring and summer. Although this is a seasonal source, a huge number of cities, from Tokyo to Seattle, tap into this freshwater, and have designed their water supplies around it. The problem is that according to the Organisation for Economic Co-operation and Development (OECD), temperatures are rising in many regions, reducing snowfall year on year and causing glaciers to shrink. As a result, the snow-reservoirs get smaller each winter, leaving a shortfall for those cities that use them to get through warm summers.

I know. It all feels a bit bleak. But this is the reality facing our growing cities. Water is the most important component that we have to consider, and it's the most under threat. We may use a diverse portfolio of water sources in order to make it more manageable, but we'll still need to find better ways of utilising the precious 2.5 per cent of water that is safe for us to drink.

Treat

Depending on the city you live in, somewhere between 4 and 10 per cent of drinking water is actually drunk, with the rest used on other stuff, like flushing toilets, washing our clothes and nourishing our plants. Few of these other processes actually need ultra-clean water, but because most suppliers treat all water equally, we do too. So how clean is clean enough?

First, despite what producers of bottled water would have you believe, truly 'pure' water is not available in supermarkets. Alongside the hydrogen and oxygen, water from any source will also contain tiny amounts of minerals such as calcium, potassium and chloride. But bottled water is a huge business. In the US, sales rose by 7 per cent from 2013 to 2014, with 50 billion litres (11 billion gallons) of bottled water consumed in those 12 months. As reported by the *Wall Street Journal*, that puts it on target to outsell soda (fizzy drinks) by 2017. These figures include plain, flavoured, vitamin-enhanced and carbonated water. My personal favourites though are the waters that claim to 'neutralise acids' and 'eliminate harmful toxins'. There is absolutely no scientific evidence to back up these claims, but when did marketers ever let science get in the way? In researching his book *Bottled and Sold*, Peter Gleick found that at least two US beverage giants get their water from municipal sources, *i.e.* the stuff that comes out of your taps. So, maybe bear that in mind the next time someone says that the bottled stuff 'just tastes better'.

Across the world, standards are in place to ensure that water – bottled or tap – is safe to drink. In many cities though, these standards are treated as aspirational, rather than legally binding, and contaminated water still kills hundreds of thousands of people every year. But let's pick up our water droplet again, and see what processes it should go through to make it clean enough to emerge from your tap.* If our droplet comes from a river or a lake, it must first survive four gruelling challenges:

1. **Coagulation** – general removal of big stuff (leaves *etc.*) and the addition of a chemical that causes tiny contaminant particles to stick together into clumps called flocs.
2. **Flocculation** – a gentle mixing of the water to encourage flocs to come together to form even bigger flocs.†
3. **Sedimentation** – the now heavy flocs of material drop to the bottom of a tank where they form a sludge that is then pumped out.
4. **Filtration** – the water is passed through several layers of sand and gravel to remove any small particles remaining, and the filtered water is removed.

If you thought that last step sounded suspiciously like the process that happens naturally in aquifers, pat yourself on the

---

* I should say that not all cities use identical treatment processes, but these steps are pretty common to all. Leaks and contamination are still a huge problem, as shown in the 2016 water crisis in Flint, Michigan.
† Flocculation is my new favourite word. Must be said with gusto.

back! Groundwater is naturally filtered, meaning it gets to skip the treatment queue.

Step five for our filtered droplet is ozone gas. Ozone (yes, like the layer) is an unstable form of oxygen – it has three atoms stuck together (O_3) instead of the normal two (O_2). Because of this, it's particularly good at breaking up other molecules, including bacteria, making it an excellent disinfectant. Tiny quantities are added to water to kill any bacterial nasties, with the excess either removed or converted back into normal oxygen.

Next, our water molecule faces its greatest test: activated carbon filtering, where it passes through a thick layer of fluffy, porous carbon particles. Here, the treatment is all about surface area. When you bake a potato, you instinctively know that cutting it into quarters will make it cook faster. That's because you expose more of the potato's surface to the heat of the oven, speeding up the chemical reaction (the cooking, in this case). This is true for carbon filtering too – just 3g (0.1oz) of these carbon particles have a surface area equivalent to the pitch in Barcelona's Nou Camp stadium. This gives the carbon plenty of opportunity to remove impurities from the water. Micro-pollutants such as pesticides get trapped in the fluff, and the ultra-clean water passes through.

Many cities take an additional disinfection step at this stage, adding minute amounts of chlorine to the water, to prevent the growth of harmful microorganisms while the water is stored. The level of chlorine present is very small and is tightly monitored to maintain it within safe limits. One thing that chlorine can't kill is *cryptosporidium*, a parasite that can cause severe and, in some cases, life-threatening diarrhoea. So these days many urban water suppliers finish with ultrafiltration, where the water is passed through membranes with holes so small that the parasite can't fit through (more on this later). Then, and only then, does water make its way to your taps.

Consume

Now that our droplet of clean water is inside our urban home, there are a number of places that it can be used. In all cities, the bathroom is the biggest drain on our water supply, with its showers, baths and flushing toilets accounting for almost half of all domestic water. Laundry comes in second place for most cities, followed by the kitchen, where water is used for cooking, cleaning, washing and drinking.

But there are other hidden users of water in your home – the clothes you wear, the energy you use and the food you eat. This virtual water is completely invisible to us because it's used long before it reaches our homes. In 2013, *National Geographic* announced that it takes around 2,700 litres (almost 600 gallons) of water to make a single cotton shirt, which would be enough water to quench a person's thirst every day for two and a half years. Before you all now panic and stop wearing cotton, know that this is an area with lots of questions, and very few straightforward answers. In 2010, DEFRA (the UK's Department of Environment, Food and Rural Affairs) commissioned a study into the environmental impact of natural and synthetic fibres. Cotton production was found to use considerably more water, but produce fewer greenhouse gases, than polyester (a product of the petrochemical industry). Wool produced far more wastewater than either of the other fibres, but used less energy than both. And bamboo produced one of the lowest levels of greenhouse gases. But, much as I would love it to be otherwise, there is no single 'wonder textile' that we can look to – all impact the environment in different ways, and water usage is a key part of that.

And what about food production? A series of remarkable scientific papers supported by the UNESCO Institute for Water Education analysed the use of water in agriculture. They showed that farmland irrigation now uses close to 70 per cent of all freshwater set aside for human use, and this doesn't include all of the water needs of livestock. The same research also looked at the water cost of individual animal and crop-derived products, and the numbers might shock you. Oils such as sesame and castor oil are some of the worst offenders, using over 24,000kg of water for every kilogram of oil produced. And this is similar to the water use in beef production (across the lifetime of the cattle). In devastating news for chocoholics everywhere, including me, the data shows that, on average, producing 1kg of chocolate (from cocoa bean to final product) takes almost 18,000kg of freshwater, and coffee is similarly water-hungry.[*]

OK, it's time to don your sceptic's hats! As we'll remind ourselves throughout *SATC*, numbers like these need to be questioned. As your friendly science guide, I'm here to help you figure out where they came from and whether they're trustworthy, so let's dig a bit. When looking solely at crops, the researchers considered a rather impressive range of factors, including average temperatures, soil quality and moisture. They also looked at the average growth patterns of plants, the rate at which water evaporates from them, the use of fertiliser and the average harvest yield (how much planted versus how much harvested). The water consumption of various processing steps, be it coffee-bean roasting or oil refining, have been well documented elsewhere, so this data too was included in the final calculations. Then all of this was averaged out across the globe. So, yes, the final figure is an estimate, but one that has been painstakingly pieced together from solid data. I'd consider it to be pretty reliable, and more than a bit scary.

Vegetables such as potatoes, tomatoes and cabbage were found to be far less water-intensive to produce than meat products, leading many to suggest that vegetarianism is 'the environmental choice'.[†] Regardless of your position on that, I hope you're now beginning to see how many ways we use and misuse water today. If predictions for future global population growth are realised, that tiny water droplet we started with will become ever more precious. According

---

[*] This is equivalent to 18,000 litres, because 1 litre of water weighs exactly 1 kilogram. This is not a coincidence, it is a hangover from the earliest definitions of the kilogram.
[†] We'll come back to the question of food security and vegetarianism in Chapter 7.

to the Institution of Mechanical Engineers, the water requirements just to meet the growing food demand could be three times higher than today's *total* freshwater usage. We will need a lot of great science and some seriously clever technologies if we're going to meet that challenge. But more than that, cities and urban regions will require a huge cultural change when it comes to food – it's not just water that we're a bit trigger-happy with! Depending on which report you read, somewhere between 30 and 50 per cent (1.2–2 billion tonnes) of all food produced never reaches a human stomach.

Food waste happens at every stage of the process: crops can fail because of bad weather or poor soil, and animals can get sick. Land usage is an issue too; livestock needs significantly more land to develop than crops need to grow. During the harvest, knobbly carrots and heart-shaped potatoes don't pass the beauty pageant standards we've set for our supermarket vegetables, so far too many are thrown away. Poor storage and transport see yet more losses, and that's all long before you forget about that sad-looking orange in your fruit bowl. There's a whole lot of waste going on, and for our cities to become truly sustainable places to live, we need better ways of balancing stuff in versus stuff out.

Out

If we're going to talk about *stuff out*, there's really only one place we can start – the loo, lavatory, WC, dunny or bathroom. Call it what you will, but if there was a prize for the single technology that made cities liveable, the flush toilet would have to win it.* The inventor of the first toilet is still a matter of debate – there's evidence that, almost 4,000 years ago, the cities of the Indus Valley (modern-day Pakistan) had a primitive flush toilet in every house, and we know the Romans later used a similar process. In 1596 an English

---

* In doing some early reading for this chapter, I found that the Wikipedia entry on the 'flush toilet' includes an audio file of the sound of flushing. Sometimes, I love the internet.

courtier called Sir John Harington published a paper on a forerunner to the modern toilet, which he called *ajax*.[*] Installed in his home, it used a valve to eject the excreta (urine and faeces) and released water to wash out the bowl. The name most commonly linked to the toilet is Thomas Crapper, but he came along much later. While we know for sure that he didn't invent the toilet, Crapper did invent the ballcock (the floating valve found in the tank), and was relentless in his promotion of sanitary plumbing practices. From Thomas Twyford, we got the now familiar single-piece ceramic flush toilet.

Historic segue completed, it's time to carry on with our water droplet's journey. You've overindulged a bit. You've had a few too many coffees and more food than you'd care to admit. Your digestive system is working overtime, but the solid and liquid waste needs somewhere to go. Depending on what city you're in, you'll either sit on or squat over a ceramic bowl in order to use the toilet. Once the excreta has... exited... and the toilet has been flushed, your faeces and urine enter the city sewers. You don't need me to tell you that sewers aren't sexy, but without them, our cities would never have been able to grow.

Knowing very little about how you build a sewer, I had to call in some help. Rory Mortimore is a retired Professor of Engineering Geology, who has worked on huge engineering projects like the construction of the Lee Tunnel – London's latest super-sewer. Mortimore told me that the first thing you need to define is a start and an end point to your sewer, and from there, the geology begins. Boreholes must be drilled along the proposed route, so that you can define exactly what rock type the sewer will eventually pass through. You also need to consider the steepness (or gradient) of the route. Sewage isn't pumped from homes and businesses to the final treatment plant, it *flows*. Given that London

---

[*] So many toilet nicknames! Ajax is probably the source of the term 'jacks', and 'john' may even be a direct reference to Harington's forename.

sewers manage 1.25 billion kilograms (1.23 million tons) of poo a year, it's vital that it flows the right way, so we make gravity our friend, by ensuring the start point is higher than the end point. The potential impact on any existing infrastructure, or even hazards such as unexploded bombs from times of war, must also be accounted for. Many rock samples and a lot of data later, the first ground plan can be drawn up. And that's just step one – the detailed planning and surveying of a major sewer system can take years. I for one am totally OK with that, when you consider the possible implications of a poorly designed sewer.

Sewage, or its more modern equivalent, wastewater, is the catch-all term for the water-carried waste removed from homes and businesses. As part of this, there's greywater – which includes any cleaning products used in baths, showers, sinks, dishwashers and washing machines. Then there's blackwater – a sinister-sounding material that includes everything flushed down the toilet, from urine and faeces to whatever you use to clean yourself (water if you use a bidet, toilet paper if you don't).

The major difference between these types of wastewater is that only the black stuff has been in contact with faecal matter. Poo contains a lot of bacteria and many pathogens, and if they find a way back into the human digestive system (e.g. through poor sanitation), disease can spread rapidly. Greywater, which hasn't been in contact with faeces, comes with far fewer health risks, and can find limited reuse within the home. In many cities across Sweden, California, Spain, Germany and Israel, to name a few, this is exactly what happens. Greywater is gathered in a separate system, and treated locally before being reused, typically within the same building. But, right now, these countries are in the minority. A UN-backed research study from 2013 showed that a third of countries don't even measure how much wastewater they produce! So as a result, most of the world's major cities send every drop of wastewater, both black and grey, to centralised plants for treatment. While the main role of sewers is to keep humans away from their poo, we make them carry far more than that.* This is a costly and inefficient way to manage struggling water supplies, and one that future cities will need to move away from.

For now at least, all wastewater is treated similarly to drinking water, before being released back into our waterways. One of the questions I was asked when writing *SATC* was: **why do we pump cleaned wastewater back into rivers?** Well, firstly, because it's all part of the same cycle – the cleaned wastewater won't damage the waterways and it frees it up for further use and treatment. And secondly, wastewater treatment really is just a faster version of what happens naturally 'in the wild' – filtering water through rocks gets rid of the big bits of gunk and bacteria in rivers breaks down pollutants. But with demands on water as high as they are, we simply can't wait for nature, so we take a shortcut using chemistry.

There is one pretty cool technology that, while originally developed for medical use, has taken the wastewater world by storm: UV cleaning. Ultraviolet (UV) light is invisible to the human eye, but is an important component of sunlight. If you've ever bought sunscreen, you'll know there are two types of UV radiation, UV-A and UV-B, which can damage our skin cells. But if damage is what you're really after, UV-C should be your first choice.† Identified in the early 1900s as an effective way to sterilise bacteria, UV-C is a mutagen. This means it can penetrate into living cells, breaking molecular bonds in their DNA, rendering them unable to reproduce. While it sounds rather vicious, this property of UV light is

---

* An important side note to this is that if you're lucky enough to live in a developed nation, your waste is at least carried away in covered sewers. In some developing nations, open sewers remain commonplace, leading to widespread cross-contamination of the water supply.

† Even in its severely depleted state, the ozone layer blocks UV-C from reaching the Earth's surface, so we don't have to worry about it.

incredibly useful – just by shining on wastewater, it can neutralise the veritable zoo of microorganisms present.

Right now, this tech is very expensive, but a number of cities have invested in it, including Moscow. In 2012, they opened one of the world's largest UV plants, which every day treats over three billion litres (660 million gallons) of wastewater.* The water first goes through the four-step process we talked about earlier. Then, it is passed into vats filled with a huge array of high-intensity UV lamps, which break down the contaminants. The UV-cleaned water can then be safely discharged into Moscow's lakes and rivers. This approach can also be used as the final step in drinking-water treatment – in fact, almost nine million New York residents have light to thank for their clean water. The Catskill–Delaware facility on the outskirts of NYC is colossal. It employs almost 12,000 UV lamps to treat nine billion litres (two billion gallons) of water per day; enough to fill over 3,000 swimming pools.

Better treatment of wastewater is one part of a sustainable water management plan, but an even bigger part is its reuse. Sydney produces enough wastewater to fill its famous harbour every year, so recycling this precious commodity has become a no-brainer. There are lots of challenges remaining, including high costs and poorly defined regulations, but the future of reusing water looks really bright, as we'll discover in our fast-forward to the future.

Fill

Not all of our 'stuff out' is quite so easy to break down. Under the streets of a growing number of cities resides a beast of terrible proportions, causing devastation wherever it goes. It is the fatberg. Comprised of fats, oils and grease, along with a charming collection of other sanitary items like wet wipes, fatbergs are a very modern problem; a direct result of our love of disposable products and eating out. Take your average bit of cooking oil. When you're cleaning up, what do you do

---

* Assuming we say that a billion is a thousand million = one followed by nine zeroes (1,000,000,000).

with it? If the answer is 'I pour it down the sink', then you have contributed to a fatberg. The problem is that the sewers aren't a magic portal – the stuff you flush down the loo or pour down the sink doesn't just disappear. Hot cooking oil might be liquid when it goes down the plughole, but when it hits the cold sewers, it solidifies, causing blockages. The growing use of wet wipes makes it worse. Even if they advertise themselves as flushable, wet wipes don't degrade in the sewers. In fact, they attract the oils floating on top of the wastewater, gradually clumping together to form thick layers of congealed fats and debris.

As you can imagine, fatbergs cause huge problems for those managing the sewers. In 2013, one was found in south-west London that weighed 15 tonnes – that's several tonnes heavier than a double-decker bus!* In Brisbane, Australia, 120 tonnes of wet wipes are removed from the sewers every year. The *New York Times* estimated that clearing 'grease backups' cost the city \$4.65 million (£3.35 million) in 2014. The solution is pretty straightforward though: don't put wipes down the loo – even the ones that say you can. And store your used cooking oil in a vessel that won't melt or break, and dispose of it in the rubbish bin.

Of course, there's a lot of waste that never enters the sewer system at all, and much of that goes to landfill.† There, bulldozers and compactors are used to squash the waste into as small a space as possible, before being covered with soil. Over time, this forces the oxygen from the soil, leaving landfills as a delightful home for tiny, oxygen-hating microbes called methanogens. These feed on any

---

* If we take the mass of one of London's sleek new Routemaster buses = 12.65 tonnes.

† The world's mountain of e-waste (gadgets, fridges, *etc.*) has been estimated to weigh in at a staggering 41.8 million tonnes – equivalent to 3.3 MILLION double-decker buses. But they could contain a literal goldmine: 1g of gold can be found in a pile of 40 phones. This is the same amount you'd get from mining **one tonne of ore**. More in Chapter 7.

carbon present and produce methane (CH_4), a particularly potent gas, best known as the main ingredient of cow farts. A recent research paper from a group of Yale scientists showed that landfills represent one of the largest sources of methane in the US, 'approximately 18 per cent of domestic emissions'. Carbon dioxide is also present at landfills, along with trace quantities of other gases. Looking at over 1,200 landfill sites, the researchers found that American homes, businesses and industrial sites send more than twice as much waste to landfill as previously thought, leading to higher methane production than predicted. All landfills across the EU and the US are required to collect this methane, but debate rages on around what should be done with it once it's collected.

This is a very typical problem for anything related to water and waste. Today's options have shown themselves to be unsustainable, at best. Existing water systems built for small urban populations are now groaning under the increased volume of water. And we behave as if there is a magical place that all waste goes to. Will the future offer any practical alternatives?

Tomorrow

In our increasingly urbanised world, the challenges of matching a clean water supply with an efficient use of waste will need to become both more focused and more diverse. For waste, we need to totally reinvent the reduce-reuse-recycle approach, not just at the scale of a city, but within our own neighbourhoods too.

The world's weather patterns bring with them a different question; one that's less about plumbing, and more about detective work – in the future, where will we get our water from? We know our climate is changing, and that warming temperatures, rising sea levels and shrinking glaciers are much more than just a vague threat. With such a tiny proportion of the Earth's water available to us, are there ways we can ensure our taps keep running? In this section, we're going to meet the vital technologies, engineering approaches and scientific breakthroughs that will change the way we perceive, use and reuse our water and waste.

Suck

Recent years have seen cities rise up from some of the driest regions in the world, from 'new' cities like Dubai, to the continued growth of Lima, Peru. The past decade has also seen a number of extended droughts, including across California and Australia. Where the demands on water are greatest, researchers are looking to technologies that suck it out of almost anything.

Air wells are structures that, without any power, harvest water directly from the air. They have been around for over a hundred years, and may even have been invented by ancient civilisations. But changing weather patterns and an in-depth understanding of materials have brought them back into focus. The key to this water-sucking technology is to use a hydrophobic material. The word literally translates as 'water-fearing', which may seem a bit weird when water is what you want. The thing is that, if you design it correctly, this kind of surface can redirect water to where you need it to go. The patterned wings of the Namib Beetle are a prime example of this. By angling its wings into the wind, the beetle can channel water along its hydrophobic or hydrophilic (water-loving) patterns, so that it reaches its mouth. A team from the Massachusetts Institute of Technology (MIT) used this idea to create a new fog-sucking net. They adjusted the diameter of the net's fibres, as well as the gaps between them, to create a series of mixed surfaces. They also worked with Harvard, who developed a super-slippery coating inspired by a tropical carnivorous plant. When bugs land on the plant's surface, they slide right into the bulb and are digested, *Little Shop of Horrors* style! The combination of this coating and the clever structure led to a net that can capture five times more fog than existing systems. It has already been tested in Santiago, Chile, and further test systems are planned for similarly dry cities.

There is also a way to harvest some of the water on Earth that we can't drink – the oceans. We've all heard the saying, 'water, water everywhere, and not a drop to drink', but there's a whole lot of science behind it. Salt water does odd things to our bodies, and it starts with a process called osmosis that happens in our cells. Cells love to be in balance, and the membrane that surrounds each one controls the flow of molecules in and out of it. Now, let's take a gulp of water from the tap – it is a pretty similar composition to the water inside our cells, so any movement of molecules through the membrane pretty much balances out. When we drink seawater, it's a different situation, because the water outside the cell has a much higher concentration of salt than that inside the cell. In a vain attempt to restore balance, water is driven out of the cell, which makes your kidneys kick in. They try to flush out the excess salt by making you pee a lot, but they can't keep up with salty seawater. If you keep drinking it, this dehydration will eventually kill you.

To have any chance of making seawater drinkable, we need to get rid of the salt. Although we talked about it in cells, osmosis happens anywhere you have two different liquids separated by a thin membrane. If the concentration of one liquid is different to the other, you get a natural flow of molecules in one direction. By putting the seawater under a lot of pressure, equivalent to 40 times higher than the recommended pressure of your car's tyres, we can force osmosis to reverse direction, leaving the salt on one side of the filter and salt-free water on the other.*

Desalination is the most expensive way to clean water right now, as you need huge amounts of electricity to run these pumps, which has an environmental impact and involves considerable cost. As a result, it's used as an emergency backup. Mostly. Saudi Arabia, a country with little or no surface water, depends on seawater desalination to produce three billion

---

* Assuming most road tyres require around 2Bar of pressure, and reverse osmosis needs about 80Bar.

litres of drinking water a day.* For now, Europe, the US and much of Asia have a choice, with other sources of water available. But, with demand for water skyrocketing, and rainfall becoming ever harder to predict, we may find that more urban areas begin to look to desalination.

Thankfully, there is ongoing research into trying to make it more sustainable. One option is to power the whole plant with solar panels. Another, more futuristic option may be to use graphene membranes. These sheets of carbon can be designed to have a precise pattern of pores, allowing water to pass through, while stopping any impurities (e.g. salt). Because the graphene is so much thinner than current reverse osmosis membranes, lower pressures would be needed to force the water through, substantially reducing the power demands. The theory was first touted by MIT in 2012, and they've continued to develop it since, producing their porous graphene sheets in 2014. This work is at a very early stage – the researchers are yet to make a small-scale working device – but it shows great promise. They're not alone either; defence contractor Lockheed Martin recently patented a water purification nanofilter, and in 2017, the University of Manchester reported on their development of 'practical' graphene oxide membranes. I'll definitely be keeping my eye on this area.

Life

Cleaning water on an everyday basis is one thing, but reacting to a crisis offers a different challenge. Rewind to 2010 and the Deepwater Horizon oil spill. It not only polluted waterways many kilometres from the rig itself, but even in 2016 much of the native wildlife is still recovering from its effects. Studies on living dolphins in late 2013 found many to be suffering with disorders known to be connected to exposure to oil. There are also countless stories of groundwater becoming contaminated thanks to poorly regulated urban industries. Given that all of our water is part of a planet-wide cycle, finding ways to rapidly clean up spillages will be vital in future.

---

* That's more than 790 million (US) gallons – enough to fill 1,200 Olympic-sized swimming pools.

The scientific literature is full of promising materials that can absorb pollutants, so I've just picked out a couple that caught my eye. China's Xiamen University is working on something that combines two superstar materials: graphene and aerogels. Graphene is very tricky to produce on a large scale, but Xiamen researchers realised that by adding small quantities of it to an adhesive and heating it to 800°C (1,470°F), it could form an aerogel – a super-lightweight solid gel material full of holes. Pump oil and diesel were no match for it; the graphene sponge could absorb up to 156 times its own weight in pollutants.

Another group, this time in Australia's Deakin University, looked to a different, and much cheaper material, boron nitride. What looks like a white powder is actually made from billions of tiny, hole-filled sheets of boron nitride. Scattered onto the surface of a spill, the powder swells to absorb it, but repels the water around it. It sucks up everything from ketchup to engine oil, and does so within minutes. Boron nitride has the added bonus of being recyclable too. Once it's done its job, the contaminated powder can be removed, cleaned by heating and then used again.

Damage

No matter what city you're in, the ground beneath your feet is a maze of tunnels, pipes and cables. As our urban jungles get denser, the demands on all of our utilities will only grow, so identifying any problems as early as possible will become increasingly important. Something we might be able to tackle in the future is the much-maligned leaky pipe. We're moving away from metals that can corrode, and towards concrete and polymers, but even in these, cracks can still form in certain conditions. Self-healing polymers that can detect and repair damage would certainly be able to help, but they've always been a bit of a pipe-dream (ha!) rather than a practical option. And this is because the chemistry is very tricky.

If you wanted to create a self-healing plumbing system, you'd first need a hard-wearing polymer to form the pipes.

Then, embedded into it, you'd need a second 'repairing' polymer that flows out only when the pipe is damaged – typically, this is stored in tiny capsules that break when a crack begins to form. You then need the liquid polymer to fill the entire crack, before hardening at just the right time. Not easy. There are countless papers on self-healing polymers, but so far, the ones that look most promising still need a bit of help – there either has to be a change in temperature to speed up the repair stage, or the two sides of the crack have to be pushed together. We are definitely getting there though, and I'd expect to see some major breakthroughs in this area in the coming years.

Another way we can look after our pipes is through structural health monitoring. This involves the use of many sensors embedded along the pipe that constantly monitor it for degradation or damage. Although sensors have long been used in bridges, this latest generation of sensor is an entirely new breed – small, wireless and low-cost – and they're gradually making their way into the water supply. Singapore is an island city-state, with limited access to water and a growing population. Despite (or maybe because of) this, they're leading the way on using sensors to manage it. Designed by a spin-out from MIT, Singapore's sensors work by continuously measuring pressure in water pipes. Because water is generally moved around at well-defined flow rates, the pressure in a pipe should be pretty constant, so any damage will show up as a pressure drop. The city has embedded hundreds of these sensors in its water pipes, measuring pressure 250 times a second, 24 hours a day. They make it possible to pinpoint the location of any leaks in real time, allowing the water suppliers to switch off the supply and minimise losses. Similar sensor networks are now being rolled out across Asia and Australasia, so we can expect the rest of the world to catch up soon. Of course, pressure sensors aren't the only option. Other researchers are developing robust, embedded sensors that can measure the pH, temperature and salt content of water in real time, to identify water contamination as soon as it begins.

As well as swarms of sensors, we're seeing rapid improvements in robotics and drone technology. Monitoring a water network is not easy – switching off the water supply just to do inspections is not the 'done' thing, and even if it were, pipes are often much too small to be inspected by humans anyway. In the case of (generally larger) sewers and waste pipes, there is another issue – air quality. Sewers are filled with decomposing food waste, faeces and other lovely things, so everything from methane to carbon monoxide can be found down there. This renders them too dangerous for regular manual inspection. Sian Thomas from London's Tideway Tunnel project believes that 'drones and robots are in, and will become ever more popular as urban plumbing systems get larger and more complex'. Floating test platforms and waterproof tunnel-crawler robots are on their way to market as I write.

One area already using drones is gas detection. Adapted from the gas sensors developed for the Mars Curiosity Rover, the latest prototype from NASA's Jet Propulsion Lab is a miniaturised tuneable laser spectrometer, weighing less than a can of soup. This sensor is looking for methane, so its laser is a very specific wavelength of light. When the laser light passes through a mystery gas, the molecules of the gas absorb energy from it and vibrate, producing a specific spectrum; if methane is there, we'll see its fingerprint. Similarly tiny sensors that can 'sniff out' carbon monoxide and carbon dioxide are also in the pipeline. These sensors are currently being trialled on a lightweight drone, and the hope is that it can be used to monitor waste streams and landfills.

Local

If there is one mega-trend for future water use, it is a move away from the centralised municipal waste treatment we currently depend on. We will instead favour 'point-of-use' treatment, where waste and water will be managed locally. Currently, around a third of all drinking water used in an urban home goes down the loo. But, with some forward planning and a bit of investment, future buildings could instead utilise recycled greywater to do the job. Many people informally reuse greywater already, collecting excess water from the shower to irrigate their plants. And a growing number of cities now provide financial support to residents who want to install their own greywater management system. However, I think we'll soon reach a stage when reusing the 'mostly clean' water from washing and bathing becomes the standard, making our water-hungry loos a little less wasteful. Commercial systems already exist, but scientists are still looking at ways to further improve the process, and to get more out of it than just greywater.

In order to be reused safely, greywater is filtered and then disinfected using chlorine or something similar. Researchers from the Cardiff Catalysis Institute (CCI) are taking another approach. They're using tiny beads of gold and palladium to transform greywater into its own disinfectant. 'As we see it, it's a three-step process,' said Professor Stan Golunski. 'We first filter the greywater as usual, and then a small amount of it is split into hydrogen and oxygen using electrolysis. This goes into a catalytic reactor containing our gold-palladium beads, and the hydrogen and oxygen recombine to form dilute hydrogen peroxide.'

Now, there is a lot of terminology in there, so let's go through it in stages. We found out about electrolysis in Chapter 2 – where water was ripped apart by electricity, leaving behind a collection of oxygen molecules (O_2) and hydrogen molecules (H_2). Eventually, these molecules would naturally find each other again and react to produce water. But in the presence of metal particles called catalysts, the reaction is speeded up, and so instead of producing H_2O, some combine to produce H_2O_2, or hydrogen peroxide. This is a disinfectant, and when used in this very, very dilute form, it can kill the bacteria found in greywater. And although the metals are expensive, they're not actually consumed in this process – they just sit there and interfere in the relationship, like a nosy relative. This means that they can last for many hundreds of treatments before they need to be replaced. So how far away are we from seeing this in our urban homes? Quite a way, I'm afraid. 'We are currently at the proof-of-concept stage,' said Golunski, 'but eventually, we hope that this could be a standard "white good" in tomorrow's kitchen.'

Local greywater treatment may become commonplace, but there are also ways to treat and reuse blackwater, yes, the waste that contains poo. Systems called membrane bioreactors (MBRs) have been installed in some of the latest residential and office buildings in Sydney, New York and Singapore – cities that all have their own water issues. MBRs work by combining two things: the insatiable hunger of bacteria, and polymer membranes filled with tiny pores. The first step is the bioreactor (also called a digester) – here, oxygen is added to blackwater, along with a special blend of bacteria. The whole thing is continuously mixed to give the bacteria the best chance of meeting and digesting the faeces. Then, the mix is passed through membranes arranged either as flat sheets or as densely packed bundles of hollow fibres.

The pores in the membranes are so small (around 100 times smaller than a red blood cell) that none of the contaminants can pass through them.* Any solid materials or bacteria in the blackwater are left on one side of the membrane, while on the other, much cleaner water emerges. The membranes require constant maintenance to make sure the pores aren't clogged, but even a small system in a building's basement can easily convert blackwater into greywater. One Sydney high-rise office not only manages the blackwater produced by the building's toilets, it also draws raw sewage from the municipal sewers and treats it too. After the wastewater goes through the membrane bioreactor, UV light is used to disinfect it, resulting in water clean enough for use in repeated toilet flushing and in the building's cooling towers.

You can also clean blackwater so thoroughly that you could drink it – the so-called 'toilet-to-tap' process. Now,

---

* The membrane pores are no larger than 0.1 μm wide. This means that the aforementioned *cryptosporidium* is about 30–60 times larger than the pores.

before you say 'eeeuw' (and I know I did), I should tell you that it already happens, and it could be the key to replenishing struggling aquifers. In Orange County, California, reverse osmosis (yes, the same process used to remove salt from seawater) breaks down all the nasties in blackwater, including harmful bacteria. After further treatment, the water is sent to the underground aquifer for natural filtering, and this is where the county gets three-quarters of its drinking water from. Reverse osmosis is just one step in a long treatment chain, so it's less 'toilet-to-tap' and more 'toilet-through-lots-of-clever-technology-and-an-aquifer-to-tap'. Because the water has been so carefully treated, it is at least as clean as the bottled water we all spend too much money on. This is not a DIY project though – processing blackwater is a complex and costly process, and it's not one you want to get wrong. Companies such as Mitsubishi, Siemens and General Electric have produced bioreactors and reverse osmosis systems that can be used to treat water at the scale of a city. Dubai, São Paulo and Munich are three major cities that have recently invested in this technology, but arguably, it's Stockholm that's leading the way. In 2015, construction began on what will be the largest MBR plant in the world. Buried deep underground in 18km (11 miles) of granite caves and tunnels, this enormous plant already cleans the wastewater of approximately one million people, before emptying it into the Baltic Sea. The new membrane bioreactor will speed up the cleaning process so that it can cope with Stockholm's rapidly growing population. But it's not just about *cleaning* the wastewater – it is possible to harvest the leftover sludge to produce other valuable resources, as we'll discuss soon.

At the moment, these sorts of membrane technologies are pretty damn pricey, so are currently out of reach for many. However, costs are dropping slowly and wide-ranging research efforts in universities will help. Having looked through a lot of research papers, I've found that novel pilot plants are now in operation in Tunisia, Italy, Japan, Hong Kong and Greece, to name a few. One example? Researchers in China recently published results on a bioreactor that can break down pharmaceutical micro-pollutants, including the headache drug, naproxen. Even the best commercial systems can struggle with this, because such compounds appear in minuscule quantities – if you took a sample of one billion molecules of wastewater, only two or three of them would be the pollutant. Key to their idea was a type of fungus that naturally munches away at pharmaceutical molecules. They grew this fungus on sawdust and added it to a large tank to form their bioreactor. Analysing the wastewater post-treatment showed that the target molecules were almost entirely removed, and at a very low cost. This is still very much a lab experiment, and we're far from seeing fungus in widespread use, but it shows that we're not just standing still on this stuff. Science continues to innovate and improve both MBRs and reverse osmosis. This will, in time, make these remarkable systems affordable for all, and help us to close the loop on our water and waste, one building at a time.

Recover

Another way for future cities to close the loop will be through small combined heat, power and cooling (CHPC) plants in our local urban community. We talked about CHPCs in Chapter 2, but one thing we didn't discuss was where they get their gas fuel from. Soon, the answer to that will be: sewage. We know that the decomposition of waste by microbes can produce methane gas. Most municipal water treatment plants actually *use* this idea: the sludge (the leftover solid waste) is added to a digester or bioreactor. There, in a lovely, warm environment, microbes are allowed to feast on the sludge, breaking it down and farting out methane gas in the process. This gas is perfect for use in a CHPC plant, and if we scale it down to a small community or even a tower block, we could be on to a winner. The wastewater produced by the residents could not only be used to provide toilet-flushing water, but it could even provide some of the building's power, heating and air conditioning!

As well as managing wastewater, some future buildings may even harvest clean water themselves. The star of Chapter 1, the Burj Khalifa, is home to what is called a condensate recovery system. Essentially, this involves capturing the condensate, *i.e.* the steam that condenses on the building's boilers. Official figures suggest that the Burj's recovery system could provide 35 million litres (7.7 million gallons) of freshwater a year. This sounds mightily impressive... until you realise that this equates to just 37 days of peak demand.* Lots of today's skyscrapers use a version of this technology, but one that caught my eye is the facade system used on the NBF Osaki Building in Tokyo. Dubbed BIOSKIN, it was inspired by the Japanese tradition of sprinkling roads and footpaths with water during the hot summer, an idea that stands on firm physics foundations.

Their system uses a rainwater collection tank on the roof, a simple treatment tank in the basement, and a facade made from a network of narrow aluminium channels, each surrounded with porous terracotta. Rainwater is directed from the roof to the basement, where it is filtered and cleaned. Then it is pumped up through the frame of pipes, which also form the building's railings. As the water flows around the frame, it is gradually soaked up by the sponge-like terracotta pipes, and with air temperatures rising throughout the day, this water gradually warms and evaporates from the terracotta. Turning water into vapour takes energy – heat energy – so, in order to evaporate, the rainwater has to draw in that heat from the surrounding air, cooling it (slightly) in the process. This is very similar to how the evaporation of sweat from your exercising body cools you down.

The developers claim that on the hottest days, the BIOSKIN facade cools the building's surface by 12°C and the air temperature around it by 2°C. These figures come from a computer simulation, and as far as I could find, they've yet to publish any measured data. But, even so, there is a lot of excitement around this idea, with many suggesting it could be one solution to the urban heat island effect. Our cities use a lot of heat-absorbing, non-reflecting surfaces, like concrete

---

* It's been widely reported that the tower goes through 960,000 litres (210,000 gallons) of drinking water a day.

and tarmac, while our streets are full of vehicles that produce heat. Together, these factoids mean that cities tend to be hotter than rural areas. According to the Environmental Protection Agency, the average air temperature of a city larger than one million people can be up to 3°C (37°F) warmer than its rural surroundings. Warmer buildings mean an increased load on cooling systems, a higher energy demand and higher greenhouse gas emissions. If BIOSKIN truly works and is applied to a number of buildings, it could offer a way out of that cycle. Of course, as we discovered earlier, trees could be a much cheaper option to cool our cities – let's have a bit of both!

Sludge

One of my favourite news headlines from 2015 had to be, 'Poo-Powered Bus Breaks UK's Land Speed Record'. Two things: I had no idea a land speed record for buses even existed, and the bus that now holds the record was powered (indirectly) by poo. Awesome. Anyway, it managed to reach an impressive 123.57kph (76.8mph), thanks to its liquid methane fuel, extracted from digested cow dung. There are other buses that run on waste too, and we'll talk about another favourite in Chapter 5. In Stockholm, it's the taxis that make use of waste, but their fuel comes from the sewers, rather than the farm. Every year, wastewater treatment produces 76,000 tonnes of sludge. Like the fuel used for the poo bus, some of this sludge is transformed into vehicle fuel for the city's taxis. Perhaps controversially, some of the poo-sludge is also used as a fertiliser for agriculture. No, this is absolutely not the same as just using a farm as a toilet. Once the sludge is carefully treated to remove pathogens and some of the compounds that make it smell, you're left with a nutrient-rich material called biosolids.* In an interview with *Modern Farmer* magazine, Washington State University soil scientist Craig Cogger said, 'We're not as grossed out by animal manure as we are by human poop.' He's right – there is a perception that it's just a

---

* I can't help but picture a discussion on what to call it: 'Just make it as vague as possible… don't mention human poo at all…'

bit 'ick' to reuse human faeces at all. And there are some who cite health concerns too, suggesting that treatment may not remove everything bad from sludge. So far, measurements look good, but analysis is ongoing.

For researchers at Rice University in the US, it's liquid waste that could be the key to future fuels. As we'll discover in Chapter 5, biodiesels aren't always as environmentally friendly as they're made out to be. But those that use algae – microscopic plants that thrive in the water – could be a much better option. They are incredibly productive, doubling their number in just a few hours. Not all algae can be used to produce fuel, but those that can need a specific blend of nutrients, and this is where wastewater comes in. At a treatment plant in Houston, the Rice team set up a pilot over four months. In that time, they used wastewater as a giant tank of delicious, nutritious food in which to grow algae. The tiny plants could then be harvested and transformed into biodiesel. In addition, the wastewater was 'cleaned' by the algae – the organisms consumed some of the pollutants that other processes struggle with. The positive results from the study have led to continued work between the university and the city of Houston, so it may be something that other cities adopt in the future.

A surprising (and probably my favourite) potential solution to landfill waste is the humble mealworm. Working with Stanford University, the Beijing Genomics Institute and Beihang University, Chinese researchers found that mealworms can not only eat certain types of plastic, but that they remain perfectly healthy afterwards! In a paper published in late 2015, mealworms were fed a pill-sized amount of Styrofoam every day for 10 days. The microbes in their gut broke down the plastic, releasing CO_2 (yes, mealworms fart) and leaving completely biodegradable droppings behind. Let me just repeat that… The mealworms took a material found in every landfill, and transformed it into something that naturally degrades over time! One of the researchers, Wei-Min Wu, said, 'Our findings have opened a new door to solve the global plastic pollution problem,' so they're not pulling any punches. Next, they need to investigate the possible impact on the food chain – does consuming a mealworm on a plastic-only diet have an effect on a larger animal? And what about the one that eats *that* animal? So, there are still questions to ask, but one day, a plastic-laden landfill may be seen as an all-you-can-eat buffet for mealworms.

Of course, another option that doesn't require a squad of mealworms is a general reduction in our use of landfill-fodder. Take the oh-so-fashionable single-serve coffee pod that wants to tempt us away from our friendly barista. Valued by the *Financial Times* in 2014 at around €9.7 billion (£7.6 billion, $10.6 billion), it is a booming market amongst busy urban-dwellers, but it also results in a huge amount of non-recyclable waste. Bioplastics, created from the waste products of the food and paper industries, may be one way to transform the pod. At the time of writing, most of the major players in the coffee market were investigating biodegradable coffee pods, so we shall see what happens. Just like today's standard plastics, derived from oil, bioplastics can also be designed to have different properties – some are resistant to high temperatures and can be recycled, while some break down after use. It's hoped that these materials will eventually replace all plastic packaging, and maybe even their use in pipes and tubing, but we're still a way off. 'Bioplastics are currently two to four times more expensive than conventional plastics,' said Paul Law, from Biome Bioplastics, and that's really because the work is mostly still being done in the lab. But companies are working with universities to develop this waste-to-bioplastic approach at an industrial scale, and if they manage it, we may finally wave goodbye to unsustainable, oil-based plastics. For now, maybe just take a minute to think before you throw your fancy coffee pod in the bin.

Frankly, on this topic we could all do with thinking a bit more. Especially those of us lucky enough to see clean water pouring out of our taps, or whose rubbish is carried away in a bin lorry. The bottom line is, there is no 'away'. Everything has to end up somewhere. More than any other system that we'll meet in *SATC*, water supply and waste removal are a complex cycle that rely on engineering at every step. If we don't implement better ways to manage that cycle in the near future, we will be in deep doo-doo.

CHAPTER FOUR

Way

Time to hit the open road, with the top down, the wind in your hair and the sun on your face. Road trips have been at the heart of countless songs and movies, always accompanied by dramatic vistas. It's rather telling that so few epic driving songs refer to the daily commute, or to traffic-induced rage. It seems to me that we have a love-hate relationship with roads – in our minds they symbolise freedom, but in reality, those within our cities are a source of daily stress. But whether you take to them by choice or necessity, their role in shaping our urban environment cannot be overstated. Like power and water, our roads are often viewed simply as utilities, but they are much more than that. Together with the rail network, they form the veins and arteries of a city's beating heart, moving people and goods to wherever they need to go.

In this chapter, we're going to learn more about our road network, and the many technologies that keep traffic flowing (or not, as the case may be). We'll also investigate the science behind bridge design, and as always, glance forward to the future of our roadways. So let's get started.

Today

First, we need an idea of scale. According to the CIA World Factbook, the world's road network was 64,285,009km (39,944,852 miles) long in 2013. Laid end to end, these roads would circle the Earth 1,600 times.* Numbers like these should

---

* The CIA World Factbook is maintained online by the Central Intelligence Agency in the US. It aims to 'marshal facts on every country, dependency, and geographic entity in the world', and then makes the data publicly available. Am I the only person who'd never heard of it?

always be taken with a pinch of salt, but regardless this estimate nicely demonstrates our need for infrastructure. We have criss-crossed the world with roads, both paved and unpaved, all because we love to be connected. Before we talk about building roads, can you guess which country has the longest road network? Yes, it's America, spiritual home (and actual birthplace) of the car. Some of the other countries that make the top ten include China, Russia and Canada – perhaps unsurprising, given their size. But the list also includes France; for me, a far less obvious candidate. Despite being about 27 times smaller than Russia, it has a similar-sized road network, and this is thanks to population density. An average square kilometre of Russia contains 8.4 people. In France, you'll find 120 people in that same area, and generally, where you have more people, you'll need more roads... and probably, more bridges.

Now and for ever a critical transport link, bridges can define a city's skyline – just think of San Francisco or Sydney. They can be nothing more than a few stepping stones, or long enough to span seas, but all bridges have one thing in common: they helped cities to overcome geographical hurdles, fuelling their growth. Like skyscrapers, bridges are popular fodder for 'listicles' – lists (always accompanied by pictures) that have titles that begin with 'The 10 longest/tallest' – and China dominates all of them. There is however, a hotly contested debate on where the title 'City of Bridges' should be bestowed. Berlin, Hamburg, Amsterdam, Venice and Pittsburgh have each been suggested at one stage.

The main thing to remember about these structures is that they exist to keep the heart of a city beating. Roads and bridges need to be numerous, safe, reliable and (relatively) unclogged in order to work, and while you might not have realised it, engineering and science lie behind all of those characteristics.

Build
Let's begin by building a road.

- Step 1 – choose the **route**. Although a straight line would be the ideal, in a built-up urban environment any new road will have to curve around buildings, or will need to rise above or tunnel below existing infrastructure. Regardless of which route is chosen, we'll need an experienced surveyor to come in to check the area's geology, soil conditions and water content.

- Step 2 – next, you need to decide on the best **design**, based on the likely capacity of the road and the weather conditions it might have to endure, as the maximum and minimum temperatures will limit our options.
- Step 3 – we'll also need to look at the sorts of **traffic** that will use the road. Will it be mostly cyclists, or heavy-duty trucks?
- Step 4 – once we've considered all of that, we can decide on the best **materials** for the job. While it's perfectly possible to produce roads using cobbles, rocks or concrete, none of these will be able to cope with extremes of temperatures or constant traffic.

Today's roads are generally built using layers of different materials, each with a different job to do. Composite roads aren't new – the Romans were the first to use them – but it was the invention of the rubber tyre that changed the game.

Rewind to 1887. A bearded Scottish veterinarian is spotted riding a strange-looking bike around sports fields in South Belfast. He got it into his head that the addition of an inflated tube of rubber around each wheel would make cycling much more comfortable. Soon after, he was granted a patent for the world's first pneumatic (or inflatable) tyre.* That man was John Boyd Dunlop, a name that's associated with tyres to this day. Made from latex and sulphur, this new 'vulcanised

---

* Dunlop's patent application was later retracted when it was realised that, unknown to him, another Scotsman, named Robert William Thomson, had registered the same technology almost 40 years earlier. But the latter's story of discovery has been almost forgotten by history.

rubber' was both elastic and strong. As well as making the road feel less bumpy for cyclists, tyres made the metal wheels more 'grippy', producing higher friction between the wheel and the road. It's this principle that today's car tyres depend on. Friction is a complicated beast, both the worst enemy and the greatest friend of mechanical systems, but what causes it, really?

I put this question to a lot of my scientist friends and I think the following covers it: we know that friction has something to do with how individual atoms of a material interact when they're in contact with another material. We also know that it doesn't act the same way everywhere – if you rub your hand along the grain of a piece of wood, and then go against the grain, you can feel that friction has something to do with direction. We think that quantum mechanics (the weird laws of physics that operate on the sub-atomic scale) could describe it. But, and this may surprise you, *that's about as much as we know*. It's not to say that we've given up, it's just that friction is remarkably difficult to investigate at the smallest scales. Scientists are working on it, though!

On the larger scale, the effects of friction are plain to see. A car is much easier to control on a dry road than on an icy one, and that's all to do with the frictional forces that exist between the tyres and the road surface. The higher the frictional forces, the better the grip.[*] But grip is not the sole responsibility of the tyres – the road surface has a role too, and we'll need to consider that in choosing our materials. In addition, the surface will need to withstand the stresses and strains that result from several tonnes of metal driving over it every few seconds. This is where a layered approach works really well. The general idea is to put your expensive, most hard-wearing material on the top, with other cheaper materials underneath, supporting it and spreading the load.

---

[*] Increasing friction is why tyres are heavily patterned too – the ridges and channels not only hold onto the road, but they pull water away from the road, further improving the grip.

A slice of urban roadway looks much like a slice of Victoria sponge cake (although it is substantially less tasty).

At the very bottom of our road-cake, there's the sub-grade – the exposed native soil that has been compacted by driving over it with a huge roller. Above that is a layer of strong, dry material that forms the foundation of a road. This is typically made from recycled concrete, collected from a demolition site and crushed into pieces. After another compression by the roller, the base material is laid down – another layer of crushed rock, mixed with a waste product from steel production called slag. For areas with lots of traffic, this layer is a must. Then comes the smelly stuff – the binder and surface materials, generally referred to as tarmacadam (in Europe) and asphalt concrete (everywhere else).

The 'tar' in tarmacadam can refer to any sticky black substance composed of hydrocarbons. Other similar words include bitumen, asphalt and pitch. In fact, these materials are subtly different, but I will lump them together here for ease (please forgive me, tar experts).* Similarly, the terms 'macadam' and concrete are used interchangeably, despite the fact that they're not the same thing. For this chapter, both refer to a densely packed gravel, with differing particle size.

So, choose your favourite sticky black hydrocarbon mixture to act as the binding agent. Heat it up until you get a nice, flowing consistency (depending on the binder, it'll be between 90°C and 160°C, 194°F and 320°F). Now mix in your gravel until all of the rock fragments are coated. Pour that mixture onto the base material until you have a thick layer. Roll it out as soon as you can – you want to pack those rocks in good and tight – and repeat. You've built yourself a perfect road!

No, roads could never be described as environmentally friendly. They use up huge quantities of materials – a rough estimate suggests that a stretch of road 1,000m (3281ft) long and 3.75m (12.3ft) wide contains close to 10 tonnes of

---

* From what I can gather, the terms bitumen, asphalt and pitch are interchangeable and refer to a substance that is naturally occurring. Tar (or coal tar) is produced by the destructive distillation of coal.

tar/bitumen, equivalent to the weight of a quarter of a million chocolate bars. (I know which I'd prefer.)* The process of road-building itself can have a huge impact on the local environment, by changing drainage patterns and forcing migration of wildlife. On top of that, the heavy-duty machinery used runs on fossil fuels. Roads are dirty, but as our cities continue to grow, we will inevitably build many more of them. I'm never one to stand in the way of progress, but by any metric this approach is completely unsustainable. Thankfully, after speaking to lots of experts, I feel assured that there are efforts underway to make roads a bit less damaging. I'll tell you all about them later.

OK, we've chosen our route, and we have all of the gravel and bitumen that our urban engineering hearts could desire. Let's also assume that our road will be flat, but that it'll need to bend and curve around buildings and other obstructions. All fine. To choose the correct lane-width and junction spacing, we'll need to consider the mathematical rules of geometric design. This topic could make up a book of its own, but fundamentally, it is about finding ways to combine straight lines and curves, to make driving as safe and comfortable as possible. If you live in a gridded city like New York (well, Manhattan), you'll be used to straight roads that mostly intersect at right angles. In terms of urban planning, this is the ideal, because grids make it easier to predict drivers' behaviour and to examine traffic patterns. For most cities, though, roads are an amalgamation of shapes and sizes, all with the aim of keeping traffic travelling at a constant speed. But even excellent geometry might not be enough to stop the dreaded traffic jam, as we'll soon discover.

Now that we've done all our prep, we have our materials, and are thinking about design. What's next? Well, at some point we'll need to figure out how we can control and monitor the traffic on our road. But we haven't yet considered

---

* I'm using the standard Dairy Milk chocolate bar unit of mass measurement here = 45g (1.6oz). Judge me all you want.

what happens if we meet an obstacle we can't avoid. Say, a river. In these cases, we really have only two options – over or under.*

Bridge

Some people can get a bit emotionally invested in bridges. I am one of those people, and to be honest, it can all get a bit embarrassing. I just love how they look different from all angles, how they feel underfoot, how they're built to last, and how they're a perfect combination of form and function. Don't get me wrong, I think some bridges are ugly, and some are a bit boring. But all of those standing today share common bits of engineering, and that's what we want to uncover.†

First though, a bit of context. Throughout history, rivers have had a vital role in supporting the development of civilisations. They have fed us, watered us, allowed us to travel and trade, and helped to build the modern world. And, more than occasionally, they've been a real hindrance! Rivers are a natural barrier to the movement of people and goods. We want to be close to them, but they haven't always been good for us. And haven't we all had a relationship like that? Without the means to cross these geographical barriers, our ancient cities would have developed completely differently. Boats were (and are) perfect for traversing waterways of course, but for a valley or gorge, the only real option was a long detour. We needed something else – a bridge. Since ancient times, people have crossed rivers on fallen trees, or assembled primitive bridges from logs, vines and eventually stone. But over time, bridge-building developed into a true science, defined by equations and rules.

Now, as a scientist I love to be exact, but as a human, I know that this approach can be a bit off-putting. So, instead of blinding you with every type of bridge ever made, I think

---

* In Chapter 6, we will head deep under London's streets to visit some of the most impressive urban tunnels in the world.
† Everything that's true for bridges is also true for flyovers and overpasses.

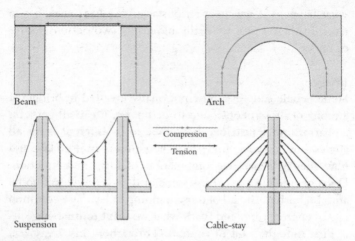

Figure 4.1 *Four basic bridge types, showing the stresses they are designed to withstand.*

we should focus only on the key features that you'll have seen on countless bridges.

We'll start with the good old-fashioned **beam bridge**. These consist of a horizontal beam on top of a pair of support pillars. If you have a couple of tins to hand, grab them – they're going to be your pillars. Place a ruler on top so that it spans both tins, and ta-dah, you have a beam bridge! You can probably see that if everything was bolted together, this would be a really good option for crossing a tiny, imaginary river. However, if you put something heavy onto the middle of your bridge, the beam (or deck) begins to bend and may eventually snap. This is still true when you scale it up, and it's because the beam is under two types of stress. The underside is stretched and pulled, thanks to tensile stress, while at the same time, the top side of the beam is being tightly squeezed – this is compressive stress. It's this constant battle between push and pull that defines how all bridges perform, how much they can support and how long they can be.

For a beam bridge, you can add more pillars to support the load, but given that we're trying to span an obstacle, more pillars may not be an option (at what point does it stop being a bridge and become a wall?). You could also make the beam from a stiffer material, or add support beams below or above the main one, so that the whole thing is less likely to bend. This approach has been widely used since the early days of the Industrial Revolution, so it definitely does work, but it only buys you a little more length. It's time for something altogether curvier: the **arch bridge**.

There's lots of evidence to suggest that engineers understood the power of the arch more than 3,000 years ago, but it was the Romans, flashy as ever, who realised its potential for creating huge stone bridges. Whereas beam bridges are all about balancing tension and compression, arch bridges really are just about compression. It's the squeezing together of components that gives this shape its strength, and it's one of the reasons arches can still be found all over the world. Key to an arch bridge's success is its (well-named) keystone – the wedge-shaped stone found bang in the centre of an arch. Because of its shape, it pushes against the stone next to it, and so on, down both sides of the arch. At the bottom you'll find the end supports, also called abutments – these are generally embedded into solid ground, which presses against them in return. This pushes the arch's component parts together even more tightly, so that the whole thing supports itself.[*] The Romans used a temporary wooden frame to support the construction of an arch – once the keystone was added, the arch was secure and the frame removed. Because these single-span stone arches depend on compression, they can't be made infinitely long. Building several arches in a row helped to extend their reach a little more, but increased the cost considerably.

The Romans rarely used solid stone for these structures. In fact, they loved a bit of cladding – they built the arch using stone, but the rest of the structure comprised compacted gravel, sand and an early form of concrete. As we entered a

---

[*] This is also true for arched windows and doorways, but instead of the abutments being embedded in the ground, the walls around the frame act as the support.

new era, we wanted strong bridges quickly and at lower cost. It was a metal that came to the rescue in 1779, when a small village in the west of England became a star. Now called Ironbridge, it was the site of the world's first iron-framed arch bridge. Still a thing of beauty, it combined compression-arch bridge design with the new lightweight wonder material. Although iron was later largely replaced by steel and other materials, the design principles for arches haven't changed at all.

The next big step for bridge design was made less than 200km (125 miles) away from Ironbridge. In the 1800s, the Menai Strait, which separates the island of Anglesey from mainland Wales, was regularly crossed by ferry, despite being a short but dangerous stretch of water. When the decision was made to construct a bridge across its narrowest section, it was clear that a multiple-arch bridge would take up too much valuable waterway. In addition, the shifting sands along the river banks could never have supported the abutments needed for an arch. Instead, engineer Thomas Telford looked to an ancient form of river crossing – the rope bridge.

Though today's **suspension bridges** look rather different from the precarious rope structure you might now be picturing, the physics is the same. In both, the roadway 'hangs' from a series of cables or ropes along the entire length, so making sure that they are pulled taut is very important. But why is it that suspension bridges can carry thousands of cars at a time, but rope bridges are a bit saggy? The difference is down to the tall towers of a suspension bridge; they give it structural stiffness by supporting the cables. On a typical suspension bridge, 'small' cables connect the bridge deck to the (u-shaped) main support cables – effectively, strings tied on to bigger strings.* The huge sweeping support cables are then suspended from the bridge's towers. Far more than an

---

* It's all relative; the suspenders on the Golden Gate Bridge are 68.3mm (2¹¹⁄₁₆in) in diameter.

aesthetic feature, these towers are the main load-bearing part
of the bridge. They turn all of the tensile stress experienced
by the cables into compression of their strong foundations,
supported by the bedrock beneath the bridge.

For me, the iconic towers of the Golden Gate Bridge are
the ultimate in suspension bridge design, so I was more than
a little excited to visit it for the first time in 2015 (honestly,
I just kept on stroking it. People stared). Despite the fact that
they support the weight of the bridge, the towers stretch
long and lean into the San Franciscan sky. Their strength is
down to two things: steel and rock. Tightly packed clusters
of hollow steel boxes make up each of the towers, making
them lightweight yet strong – think of them as bundles of
drinking straws. Importantly, this design allows them to
bend and flex (rather than buckle) under the tension from
the cables. The rock bit is less obvious. If you've ever built a
rope bridge or put up a tent, you know that finding a firm
anchor point is vital. So where do you anchor a cable that
weighs 2,000 tonnes?[*] For the Menai Bridge, Telford used
bedrock. He blasted an 18m (60ft)-long tunnel into the
rocky hills on either side of the Strait. At the end of each
tunnel, strong iron frames were installed to brace against
the rock. The ends of the main bridge cables were threaded
through the frames and secured in place by huge bolts. The
same system continues to hold the cables today, over 180
years later.

Cable–stay bridges are a little different – their cables are
tied directly to the centre of the deck, and as with a suspension
bridge, they channel the tension into its towers. They use less
cable than suspension bridges, which keeps the weight
down – important if you want to produce particularly elegant
structures. To my eyes, two of the sexiest bridges in the world
are cable-stays: the Samuel Beckett Bridge in Dublin and the
Millau Viaduct in southern France. If you don't know them,
have a little look online and thank me later.

---

[*] The 16 original chain cables that Telford used in the Menai Bridge
each weighed around 120 tonnes.

Withstand

Steel plate may be the most commonly used material in today's bridges, but as a materials nerd, it's the steel in the cables that I want to mention. From the outside, bridge cables look solid, but in fact they are made from tightly wound individual strands of steel. Each of the main cables of the Brooklyn Bridge has 9,000 strands packed inside. And we use this structure because it gives them super-strength.

If you looked at solid steel under a microscope, you'd see a mix of iron, carbon and other elements, arranged in crystals. Put it under enough tension (basically, hold one end and hang a weight off the other), and the bonds between crystals will break. But if we carefully stretch steel into fibres, its crystals line up, and the bonds between them get stronger. This means a piece of wound steel cable can withstand far more tension than a solid piece of the same size. It is a little like cotton – in its raw state it can easily be pulled apart, but when spun into a fibre, it can hold clothes together. The cables used on today's bridges are unbelievably strong – each of the main cables of the Millau Bridge could hold back 25 jumbo jets with their engines at full throttle.

There are many other things to consider when building a bridge, but just as in building skyscrapers, wind is the major one. When wind hits the side of a bridge, it forms pockets of disturbed air that change the air pressure around it. As a result, some parts of the bridge deck will be either slightly sucked up or slightly pushed down, and the cables and towers flex against the movement. This brings us back to resonance, which we first discovered in Chapter 1 – the motion of the bridge can interact with the wind and amplify itself. The famous footage of the dramatic Tacoma Narrows Bridge collapse shows just how far this twisting and flexing can go, but if a bridge has been badly designed, even a light breeze can cause it to pull itself apart.*

---

* Resonance wasn't the only factor at play here – it may have caused some of the initial motion, but it was when cables began to snap that it became self-destructive.

This is a particular problem for bridges with a thick deck or roadway. One option is to streamline the sides of the deck, to channel the wind harmlessly over and under it. Alternatively, instead of a solid support structure under a deck, you can use an open box made from steel beams, to allow the wind to blow through unimpeded. Bridge cables often get wind-proofing treatment too. Every cable on the Rio-Antirrio Bridge in Greece has a small fin wrapped around it; this stops any build-up of resonance.

Given that most bridges are made predominantly from steel, corrosion is another big problem. There are two ways to avoid it: one, you could use stainless steel, which thanks to its natural coating of chromium oxide, doesn't corrode.* But it's at least four times the price of normal steel, so that's likely to rule it out. The more commonly used second option is to paint or coat the steel, protecting it from the ravages of the wind and rain. This can't be done just once, though – on most bridges the painting never stops. The third option may seem a bit odd. You know that rusty beam bridge across a busy road that has always worried you? Well it's probably made from weathering steel, a material specifically designed to rust. To learn more about it, I spoke to my former colleague at the National Physical Laboratory, Dr Alan Turnbull (or God of Corrosion, as I like to call him). It seems that in this metal, the rust layer we see acts as a stable barrier coating to prevent oxygen and moisture from reaching the rest of the metal. Just like paint, the barrier doesn't stop rusting completely, but it can slow down the process. According to Alan, 'In certain applications, the corrosion rate of weathering steels is low enough to need only nominal maintenance, and they can attain a design lifetime of over 100 years.' This stability makes weathering steel a material of choice in highway structures and railway bridges. But they have their

---

* As an aside, this oxide is the reason that stainless steel cutlery doesn't taste metallic when you put it in your mouth. To learn more, I highly recommend *Stuff Matters* from material scientist and all-round awesome bloke, Mark Miodownik.

limitations too. 'Salt speeds up corrosion,' said Alan, 'meaning that weathering steels aren't suitable for use in the marine environment. Even poorly applied road de-icing salts could cause problems, so their use must be carefully considered.' In the right environment, these steels are an excellent choice, but whether they look good or not is a matter of opinion.

Resisting corrosion and wind are just everyday challenges for bridge designers, but in many cities, bridges must cope with something much greater. San Francisco, Osaka-Kobe, Istanbul, Mexico City and Jakarta are just some of the major cities at risk of a catastrophic earthquake. So how do you build a bridge that can withstand the ground beneath it ripping apart? To learn more, I spoke to the team behind the new San Francisco Bay Bridge. One section of the old bridge collapsed during the 1989 earthquake, but the team behind it are confident that the rebuild won't suffer the same fate. 'Our ultimate goal is to ensure the bridge stays open when – and it is when – a major earthquake hits the area,' said Brian Maroney, the chief bridge engineer. 'In an emergency, the Bay Bridge will become a lifeline.'* One of the ways that the team are attempting to future-proof the bridge is through shear link beams. Connecting the four legs of the bridge's tower, these steel beams are designed to bend and flex in the case of a quake, allowing the legs to move independently. This means that they absorb most of the seismic energy of the quake (they 'take the hit'), ultimately protecting the tower from fatal damage.

Arguably the most futuristic bridge in the world, the 4km (2.5 mile)-long Akashi Kaikyō Bridge links the city of Kobe to Iwaya in Japan. Its 100-storey towers, three times taller than those on the Golden Gate, are the first line of earthquake

---

* In 2015 the Bay Bridge team came under fire when advanced corrosion was identified on steel rods connecting the bridge's tower to its foundation. At the time of writing, $4 million (£2.8 million) had been approved for additional tests and an expert panel from the National Academies of Science and Engineering had been convened to review the work.

defence. Each contains 20 huge damping pendulums.* If a tremor causes the tower to lurch one way, the pendulums will swing in the opposite direction, counteracting the shift and preventing the tower from toppling. As well as this, other technologies are used to maintain the enormous bridge. Dry air is constantly injected into its cables to minimise corrosion, and a team of maintenance robots regularly inspect, paint and repair the bridge.

So, after all that, we have our road surface and we've chosen our favourite bridge design. What's next?

Traffic

It's happened to all of us. You're stuck in a horrible city-centre traffic jam. Everyone in the car is cranky, and every traffic light you reach seems to turn red instantly. Edging forward slowly, you find yourself asking, 'Surely it can't be that hard to manage traffic?' It turns out that there's a huge pile of maths behind understanding traffic jams. If you've ever driven on a motorway/highway, you'll know that traffic congestion doesn't always happen at junctions. Sometimes it's just because there are too many cars on not enough road, or maybe an accident or roadworks are to blame. You've probably also witnessed many phantom traffic jams too, where for no discernible reason, traffic builds up and then eases. A number of years ago, a group of Japanese physicists rented a closed circular track to investigate what would happen to traffic flow in the absence of a bottleneck. Twenty-two volunteers in different cars were instructed to get up to 30kph (just under 20mph) and maintain that speed at a safe distance from the car in front. Very quickly, the system broke down, with some cars at a standstill while others sped up.

The reason is surprisingly simple – people have trouble maintaining a constant speed. One person drives too fast, so then puts a foot on the brake to correct for it. The person behind them then over-compensates for the sudden braking,

---

* Yes, these are similar to the dampers used in earthquake-'proof' skyscrapers that we covered in Chapter 1.

and so on to the car behind. This effect continues to ripple back to other cars, growing all the time, until traffic eventually grinds to a halt. On a busy motorway, once one person brakes too hard, it can cause a start-stop 'shockwave' that travels backwards. From above, you'd see a road filled alternately with tightly packed cars and sections of busy-but-moving traffic. According to Tom Vanderbilt, author of the book *Traffic*, 'You're not driving into a traffic jam, a traffic jam is driving into you.' It's worth mentioning that mathematicians at Temple University in the US have shown that these waves (which they call jamitons)* still occur even when everyone is driving perfectly. According to Temple's Benjamin Seibold, 'we're inclined to blame individual drivers, but the models show that even if no-one does anything wrong, these waves can still arise'.

The use of maths to better understand traffic waves has been termed jamology.† It can also be used to understand how people queue, or move through train stations – regardless of the situation, we always see the same outcome. Well, almost always. Despite their huge numbers, rows of worker ants moving along bottleneck-free paths never fall victim to phantom traffic jams. Results suggest that this might be because individual ants give each other a lot of headway. This means more time to react to any incidents up ahead, making ants less likely to 'brake' harshly. There's a lesson in there for all of us drivers!

In our built-up urban centres, traffic lights are our main weapon in the war against congestion. Today, each signal houses a single coloured LED (light-emitting diode), alongside a cleverly designed piece of glass or plastic called a Fresnel lens (mentioned back in Chapter 2) to magnify the light. Instead of a smooth lens like you'd find in your glasses, Fresnel lenses are made up of lots of concentric ridges. Because the light is forced to bend around

---

* The secondary 'wave' that is produced further back along a busy road is called a **jamitino** – the baby of a jamiton. This has now been added to my list of favourite words.

† The term **jamology** was first suggested by Katsuhiro Nishinari of the University of Tokyo.

these ridges, what you see is a 'focused' beam of LED light, easy to spot even at a distance.* While the specific design of traffic lights can vary between countries, the technology that allows them to detect the presence of a car is the same the world over. Some traffic lights are set by a timer – they don't really care about traffic conditions. This is the most basic way to control traffic and can be sufficient for a road that is always heavily congested. A much better approach is vehicle actuation, where the number of vehicles defines how long each light stays green (or red). If the traffic in one road of a junction is at a standstill, the surrounding traffic signals can be automatically adjusted, to give those stuck bumper-to-bumper a little more 'green time'. While this sounds very impressive, the detection process is simple.

The most common vehicle detector depends on the close relationship between electricity and magnetism that we talked about in Chapter 2. Quick recap: if you pass an electric current through a metal wire, you induce a small magnetic field around it, and the same is true vice-versa – if you put a metal wire into a magnetic field, you induce an electric current. By wrapping a wire into a loop, you amplify this 'inductance' effect. Traffic engineers install these inductive loops in the road surface, close to the eventual position of the traffic light. When a car (typically, a big chunk of metal) drives over the loop, it changes the way that electricity flows in it. When the road is empty, or the car moves off, the coil's electrical signal returns to normal. In this way, inductive loops can constantly monitor traffic flow, counting the number of vehicles that pass by in a given time. They're easy to spot in the road too – just look for a circular or rectangular pattern of rubbery-looking stuff near the lights. But you'd better do it soon, because inductive loops are gradually being replaced with tiny, battery-powered magnetic sensors. They work in a similar way, by detecting magnetic changes in response to a car's movement, but they can also send their data wirelessly, removing the need for cables.

---

* Fresnel lenses were originally designed for use in lighthouses, but are also used in car headlamps, telephoto lenses and projectors.

Now that we have information from the sensor, we can use it to adjust the traffic light's switching pattern. For this you need a controller, pre-programmed with a series of simple options – for example, continue to increase the green time until either the traffic lessens, or there is a conflicting demand from elsewhere in the junction. But in a city with 6,000 traffic lights (like London), this probably won't be sufficient. Back in 1999, a Greek mathematician called Christos Papadimitriou officially proved that while one road junction is easy to understand, multiple junctions are a mathematical nightmare – sorting one out has a knock-on effect elsewhere. In fact, what he found was that traffic light switching in a large complex network (like a city) is what's called computationally intractable. This means that while the solution to perfect traffic light switching exists in theory, it would take so long to solve that it's not practical. In other words, it's not something that we can solve by throwing money or computing power at it. All we can do is find a compromise, an optimal switching pattern that allows traffic to flow as smoothly as possible.

To learn more about how all of this works, I visited Transport for London's (TfL) traffic control centre to speak to some of the experts. There, I watched data from London's roads stream in, and I got to control a camera at a busy junction (I was embarrassingly excited about this). After much button-clicking, I learned that the jewel in the traffic control crown is a complex software system called **SCOOT** (Split Cycle Offset Optimisation Technique), housed in TfL's supercomputer. Just like other traffic control systems, SCOOT makes use of the data from inductive loops and magnetometers to find the optimal way to switch traffic lights. As you may have guessed from its natty acronym, it does more than just increase the green-light time. SCOOT has three things that it constantly measures:

- Split: the amount of green-light time;
- Cycle: the time it takes to go from one green to the next green;

- Offset: time between green and red (the amber/ yellow-light time).

When a junction (or series of junctions) is congested, the controller in the lights doesn't just increase the split. It can change any or all of the three settings; whatever it takes to ease the traffic in a given part of the city. It also adapts all of this in real time, and does so continuously, 24 hours a day, 365 days a year.* Glynn Barton, who runs TfL's SCOOT system, is a bit of a dude. A Londoner by birth, it's clear that he loves the city, and the complexity of managing its traffic. He told me that, 'SCOOT is used in Dubai, Cape Town, Beijing, Santiago and Minneapolis, to name a few. But I'd like to think that we're absolutely pushing the limits of what it can do, and finding ways to improve it all the time.'

Barton's team at TfL consists almost entirely of engineers who are also experienced computer modellers. For them, data is king, so they don't just depend on the in-road sensors to optimise their cycles. Data from traffic cameras is also digitised and continuously scanned; when areas of no movement are detected on the road network, an alert is raised and the images checked. London's roads are rarely empty, so no movement generally means congestion. This data is all fed into SCOOT, giving it the ability to predict where problems are likely to arise. GPS data from buses is also used to optimise signal switching, and to monitor traffic build-up. As part of the SCOOT system, TfL are pioneering the use of pedestrian and cyclist monitoring. 'Traffic control is definitely not just about motor vehicles,' said Barton. 'Future London will have more people using our pavements and cycle ways than ever.' We'll learn more about that tech later, but for now, let's go for the headline: TfL say that their use of SCOOT across the city has reduced traffic-based delays by 12 per cent in the past decade.

---

* Random factoid: one year on Earth is actually 365.25 days long. That's why we add an extra day in every four years ($0.25 \times 4 = 1$).

Londoners reading that last sentence are probably thinking, 'What a load of rubbish, the roads are busier than ever.' This is true – despite the growing availability of public transport, more people drive into London than ever before. While this trend isn't reflected in every city, for many, congestion remains a huge problem. The difference is that in those cities with systems still based entirely on pre-programmed settings, the increase in traffic has made gridlock the norm. In cities using more responsive systems (like SCOOT), traffic controllers have the opportunity to change settings in real time, to allow traffic to flow as much of the time as possible. Yes it's still busy, but the traffic is moving. The other key thing to remind ourselves of is that there is no magic button that will find the perfect way to measure, monitor and optimise traffic flow. As we've discovered, traffic-light switching is a remarkably complex mathematical problem with no easy, practical answer. And we now know the power that braking by a single driver can wield on a city's traffic patterns. For as long as our roads and cars look as they do today, congestion will continue to plague our city streets.

Don't despair! In our megacity future, all trends point to autonomy – from driverless cars to self-controlled, ultra-high-speed trains. If you believe the hype, this will come with unimagined reliability, unmatched safety and a perfect, flowing, error-free service... we shall see! It's clear that today's roadways genuinely are at a crossroads. Future driverless cars will be at their best in a completely different infrastructure from those we have right now – there'll be no need for physical stop signs or traffic lights. But in the medium term (20 years or so), few cities will be able to make wholesale changes to their road network, because there will be a *mix of vehicles* on the roads – some will be driverless, and others will have humans at the wheel (more on this in Chapter 5).

For now, the priority is to find ways to drastically improve existing roads, bridges and traffic control systems, while also futureproofing for tomorrow. Many of these challenges can be solved by being smarter with our data, and in developing computer systems that actively learn from what they measure.

In this future, there is a central role for experienced traffic engineers, but perhaps not in the way you might expect.

Tomorrow

Looking forward is always tricky when it comes to infrastructure, but for me, this is especially true for the roads of the future. I guess that's because roads don't operate in isolation. They exist exclusively to connect people and places to each other, so any tech that can improve them will have implications far beyond the tarmac. Perhaps you already have an image in your mind of what roadways will look like tomorrow. Now it's time to see if it is plausible.

Surface

Roads and their construction are a filthy business, but, at the risk of contradiction by the sudden development of a cheap-and-cheerful hover car, we're still going to need them in our megacity future. Thankfully, across the globe, researchers are finding more sustainable ways to go about it, and I spoke to one of them in the city of Madhurai in southern India. Professor Rajagopalan Vasudevan is what Irish people would call a charmer. Known to most as 'The Plastic Man', this softly spoken chemist bubbles over with excitement when he talks about his work. Rubbish (or garbage) is a major environmental issue for India. Back in 2009, the nation's then environment minister, Jairam Ramesh, said, 'If there is a Nobel Prize for dirt and filth, India will win it.' These days, more than 15,000 tonnes of plastic waste is generated daily, but Vasudevan has found a low-tech way of using it – he and his team build roads.

'We can use everything, so long as it's thinner than 80μm [0.08mm],' he said. 'It doesn't matter whether it's rubbish bags or packing foams, they can all be turned into roads.' The process Vasudevan used is surprisingly simple: first, waste plastic is collected, cleaned and shredded. The gravel that will give the road its strength is added to this plastic confetti, then it is all heated to around 150°C (302°F) and mixed until every last pebble is covered in plastic. The now glossy gravel can be added to bitumen, and laid onto a prepared surface like any standard road. Vasudevan estimates that his waste plastic substitute could replace as much as 15 per cent of the bitumen on every road. The advantages of this approach are numerous – it removes plastic from landfill, it is incredibly low-cost and doesn't require any specialist equipment. Of course, by using a waste product for part of this process, you also cut down on the environmental impact of road laying. The Marshall Stability test, which determines the maximum load a surface can support, shows that plasticised roads can carry more than double the weight of standard bitumen roads. When I asked about potholes, Vasudevan said, 'We haven't seen a single one yet! We think this is because the plastic coating on the stones stops water from penetrating between them and the bitumen, and that's what usually causes cracking.'

To date, Vasudevan's plastic roads stretch for 20,000km (12,000 miles) across 11 states in India. But I doubt if he'll stop there. For me, projects like these show engineering and science at their very best. They prove that a solution to a problem doesn't always have to be sexy or shiny to be impactful. And those that solve two problems at once? Winning. Somewhat related is a project from Dutch construction firm VolkerWessels. They are using old water bottles to produce tough plastic slabs for future roads. There is very little (read: no) publicly available data on the product as yet though, so I expect they're still some way off mass-production. But a trial is planned for 2018. Looking even further ahead, French researchers are developing an asphalt made from microalgae. Results published in 2015 suggested that this algal binder behaves similarly to standard asphalt, and can cope with similar loads. Unlike Vasudevan's plasticised roads, this work is at a very early stage, with many questions still to answer. Regardless, if I were you, I'd watch this space carefully over the next few years.

Span

For bridges, one of the big trends is the 3D printing of construction materials. In 2015, a Dutch start-up called

MX3D announced plans to build the world's first 3D-printed footbridge across an Amsterdam canal. They have developed robotic 'printers' that can draw highly complex 3D structures in mid-air, using metals, such as steel and aluminium. Key elements of their technology are the welding tips of the computer-controlled robot arms. There, metals are heated to 1,500°C (2,732°F) and emitted drop-by-drop, to gradually build a metal structure. These robotic welders have made smaller structures, but this bridge will take the process to a whole new level. Now, much as I love an ambitious project, as a material scientist, I couldn't help but wonder about the bridge's mechanical strength. Could a structure made entirely from solidified blobs of melted metal really support the weight of passing pedestrians?

I wasn't so sure, so I spoke to Lindsay Chapman, a senior research scientist at the National Physical Laboratory. As well as being one of my best friends, Lindsay is also a metallurgist – she melts metals to understand their performance, and to predict their behaviour. For her, there isn't a simple answer. 'I have no doubt that given enough investment this bridge could become a reality. However, metal alloys such as steels are incredibly complex, especially when processed at high temperatures.' Chapman continued, 'It's very likely that there will be variation in the mechanical properties of the welds – some will be strong, and others weak. The challenge will be knowing, with certainty, the behaviour of the overall structure, and to be confident that it can perform in the environment.' The bridge is planned to span a canal in the city's famous red light district by 2017. It sounds like the team have a huge amount of work to do, but I'm intrigued to see the outcome!

3D printing is making waves in the world of concrete too. In 2014, Chinese construction firm Winsun announced that they'd printed and assembled 10 single-storey houses in just 24 hours.* The following year, they unveiled a five-storey

---

* They were more like garden sheds to be honest. Still cool, but definitely shed-like.

apartment building, printed using the same technique. Detailed technical information on the process is impossible to find, as demonstrated by *Guardian* journalist Nicola Davison who wrote, 'Winsun's 3D printer is 6.6m tall, 10m wide and 150m long, I'm told. I am not permitted to see it.' What we do know is that Winsun's building material is a mixture of cement, fibreglass, steel, crushed recycled rubble and binder, and that it cures in 24 hours. Similar techniques can be used to create pillars and walls for bridges and roads, but despite the media furore in China, there are still some fundamental questions to answer. Dr Richard Buswell and his team from Loughborough University have looked specifically at the mechanical performance of printed concrete. They've shown that it is possible to print large-scale parts with a similar density and strength to standard concrete. 'The challenge to turning this into a commercial process is in maintaining that quality. The larger printing that has already been demonstrated is a great example of what scale is achievable, but it is less clear how it works structurally or aesthetically.' In addition, most structural concrete relies on steel bars for reinforcement – not something that can easily be added to the printing process. Dr Buswell assures me that they're working on it though, so it seems that 3D printing will have a real, media-hype-free role to play in the cities of tomorrow.

Maintaining any concrete structure comes at a cost. Some estimate that €6 billion (£4.7 billion, $6.5 billion) is spent on the maintenance of Europe's bridges, tunnels and retaining walls (often used on roadsides). So anything that could reduce that cost would be worth investigating. One of the leaders in the field of self-healing concrete is Hendrik Jonkers from Delft University. Surprisingly, Jonkers is not a construction engineer, he's a microbiologist. His concrete is filled with cement and aggregate, as well as billions of bacteria and a calcium-based nutrient. When a crack appears and water starts to infiltrate it, the bacteria wake up and munch away at their food. What's left behind is limestone (calcium carbonate) that can gradually solidify, filling the gap. Other approaches include using hydrogels similar to those found in nappies (diapers) as a temporary filler for concrete cracks. South Korean researchers are looking at a solution based on tiny capsules of light-sensitive glue. When cracks form, the capsules break open, releasing the material, which solidifies, thanks to the sun's UV light. Cool, right? These techniques generally only fill the smallest cracks, but all have the potential to improve the lifetime of concrete. There's no doubt in my mind that we'll see much more of this work on future roads.

The sad reality is that despite its production contributing to global greenhouse gas emissions, concrete is still the most widely used material in the construction industry. Efforts like 3D printing will certainly go some way to reducing the volume of concrete we use, but longer term, we'll need something more sustainable. In bridge construction, that 'something' might be composite materials. In Chapter 1 we met carbon fibre, a composite material made from bundles of carbon threads woven into a mesh and impregnated with a polymer called epoxy. Already widely used in the aerospace and automotive industries, fibre-reinforced polymers (FRPs) like carbon fibre can be expensive to produce, but are much lighter than metals and don't suffer from corrosion. They're also famous for being ultra-strong, but it's not quite that simple. Because they're filled with fibres woven into a pattern, their strength varies with fibre alignment. Are you wearing a woven item of clothing right now? If so, give it a tug along the weave, and then at an angle to it – the fabric will stretch differently depending on the weave direction, and the same is true for FRPs. However, it's very possible to design components for bridges and road columns that make the most of this directionality.

In 2013, bridge engineers in Washington DC had to find a way to future-proof a historic bridge whose concrete deck was crumbling under the weight of traffic. They couldn't make any major changes to the 'look' of the bridge, and the busy waterways underneath it limited the thickness of the new deck. So engineers looked to composites, using an 18cm (7in) layer of glass fibre-reinforced polymer for the new deck. The weight savings were enormous – the new deck weighed less than a quarter of the one it replaced, but (and this is the main issue with composites) the cost was doubled. The engineers were quoted as saying that the savings would be made elsewhere, and that the deck 'would last twice as long' thanks to the lack of corrosion. However, this brings up another issue: many FRPs degrade when exposed to UV light, and so the race is on to find durable coatings that protect them from, effectively, sunburn. In general though, composites are in a very good place right now – we know more about them than ever, and we're beginning to see their use in civil structures such as bridges. As we try to move away from concrete, these adaptable materials may come to the fore.

Electric

So, other than structural materials, what kind of technology will we see on our roads? For electric vehicles, the next big leap will come from in-road inductive charging. Currently being trialled to charge electric buses in Gumi, South Korea, it uses a technology similar to the charging mats already available for gadgets. Embedded sensors in a seven-mile stretch of road detect the presence of an electric bus. If one arrives, magnets installed under the road surface switch on, producing a magnetic field that transmits to the bus's receiver. The receiver converts the magnetic energy into an electric current that charges the batteries. If rolled out on a larger scale, this technology could reduce the reliance on charging points, vastly improving the range of electric vehicles.

A spin-out from the University of Auckland, called Halo, is using the same process, but this time to recharge *all* electric vehicles. What's the difference, I hear you ask? Charging a fleet of buses is no mean feat, but because they are identical, they all have the same batteries, same capacity, same charging profile, and so on. Designing a single system that can charge a range of different vehicles is a much bigger task. There is no standard setup across the current range of electric vehicles – every manufacturer uses a different system. Think about having a bucket of water with 10 bottles to fill. Five are identical (they're the buses), so you know exactly how best to add the water, and how much they hold. The other five are oddly shaped, and no two are the same (each representing a different electric vehicle); filling them will require a lot of planning.

Halo's tech wants to become that one-size-fits-all option. Since being bought by a telecom giant in 2011, their charging tech has been gradually rolled out in a number of cities, with dedicated parking spots now set up to charge your car while you're out and about. Plans are also afoot to extend the reach of this technology in a similar way to the bus-charging lanes in Gumi. Definitely something for the future. The Intelligent Transportation Systems office (at the US Department of Transportation) are interested in standardisation of charging too. They're investing heavily in connected vehicles research, building a number of real-world 'labs' in cities across the US, including Orlando, Detroit and Palo Alto. Researchers and manufacturers are now using these test beds to speed up the development of tomorrow's 'smart road' infrastructure, where connected cars can communicate with (and respond to) each other and with traffic lights and other sensors.

Now, I couldn't talk about the roads of tomorrow without mentioning solar power. A small company called Solar Roadways hit the headlines a few years back with their idea of replacing asphalt surfaces in roads and carparks with rugged, hexagonal solar panels. Road markings would be replaced by LEDs in the panels, heating elements would keep them free of ice…, and they could (in theory at least) produce electricity from sunlight. Since launching, they've had financial backing from the US Department of Transportation, as well as a remarkably successful crowdfunding project.* It's a big idea based on a known technology, and it could change the world. So, how practical is it?

Sorry to be a buzzkill, but this and other similar projects pose more questions than answers. How much power can solar roads truly produce? I am pretty sure that it will be considerably less than solar panels used on rooftops. Are they

---

* Their crowdfunding campaign raised a staggering $2.2 million (£1.6 million) in 2014.

safe to drive on? If they're smooth, cars won't get any grip; if they're rough, they'll capture even less sunlight. Are they strong enough to withstand heavy traffic? How will they store the electricity they produce? How cost-effective are they? Can they be easily maintained or replaced? Look, I know I sound very negative, but I'm absolutely not a solar road hater. It's just that for now, we have very little data to back up the well-publicised claims. But that hasn't stopped the excitement around this tech. In 2016, the French government announced plans to install Wattway panels onto 1,000km (over 600 miles) of the country's roads, and in the Netherlands, a cycle path got similar treatment. Dubbed SolaRoad, their system uses concrete paving slabs embedded with solar panels. Glass is employed here too, but the researchers carried out detailed roughness tests to design a surface similar to asphalt. In the first six months, the 70m (230ft) strip generated more than 3,000kWh of electricity, about enough to power a single household for a year. So maybe there's something in it. I'll reserve judgement until I see more data (once a scientist …).

Road lighting and markings are other areas undergoing major change. A Dutch town recently replaced the streetlamps on a stretch of highway with glow-in-the-dark painted stripes. Unlike the neon paint you may have played with in school (or worn in nightclubs), this particular product glows brightly for up to eight hours. It works by a process called phosphorescence, in which some wavelengths of light are absorbed by the paint molecules, while others are reflected. The result is a paint that, when added to the road, looks very TRON-like*, so I'm a big fan of it, in theory. Glow-in-the-dark things regularly make news headlines, with plants, trees and algae-powered lamps all proposed in 2015–16. If you've ever tried finding your way around a darkened room using the dull glow of a TV on standby, you'll understand some of the issues around this general idea. Yes, the glow can help us identify the rough outline of the room, but it's in no way bright enough to see

---

*Yes, I mean the 1982 sci-fi movie (I wasn't a fan of the remake to be honest). Think dark surfaces with neon trims and you've got it.

any details or to navigate safely around it. Will the painted stripes replace streetlamps? Colour me sceptical.

Another technology being widely touted is lights that use sensors to detect the presence of vehicles on a roadway and adjust their brightness accordingly. While this wouldn't be suitable for a busy road, it could work well for those on the edges of towns and cities. With so many of the most-hyped future lighting technologies involving decreasing light levels, I started thinking about the effect that this might have on safety. A 2012 study carried out in New Zealand showed that brightening streetlights by a small amount decreased the number of accidents of all kinds, from minor intersection crashes to those involving pedestrians. So how can we balance out the safety benefits from having brighter lights, with the need to reduce energy usage?

It seems the answer is to move to LEDs. Already in use in many urban homes, LEDs get their light from the movement of electrons in semiconductor materials (like those we discussed in Chapter 2). Mike Jackett, the author of the NZ roads study and a lighting-for-transport expert, told me that 'LEDs have made and will continue to make big changes to the way we light our roads. They are dimmable, modular and produce white light – they're a step change technology.' The best LEDs produce bright, broad-spectrum light, which because of its similarity to daylight suits our eyes much better than the yellow sodium lights of old.

I forced Mike to speculate a bit on what the far future might look like (engineers aren't fans of speculating), and he said, 'Street lighting columns themselves are often seen as potential sites for other urban smart technology. Perhaps when the whole road network uses autonomous vehicles, the street lighting levels can be reduced down to just pedestrian security levels.' Nick Chrissos from networking giants Cisco is in complete agreement on that: 'Every city is looking for a way to produce intelligent, connected lighting. We're already looking at adjusting the lighting according to the numbers of cars on the road.' So, the future for roads looks bright!

People

Now, close your eyes and try to picture the road of tomorrow. Maybe it glows in the dark, or it's been 3D printed. Perhaps it can produce its own power or can charge up electric vehicles. No matter what your vision is, you can be sure that traffic will play an important role. But we're not just talking about managing higher numbers of cars and buses on the road. Most cities are also seeing a huge increase in the number of pedestrians and cyclists moving along their streets. So let's talk about the technologies that will help the most vulnerable city dwellers: those without engines.

Earlier on we talked about TfL's SCOOT system. On the day I interviewed Glynn Barton, he had just launched a brand-new project especially for cyclists. For years, London papers have been filled with tragic news of cyclists being killed on the roads of the city, so Barton and his team are hopeful that this technology will help to keep more people safe. Using a combination of radar and thermal imaging, Cycle-SCOOT can detect riders as they approach a junction, and where high numbers are present, it can adjust the traffic signal, to give them more green time. Although it's only at the trial stage, if it goes well, it will give London cyclists an official head-start from the traffic lights.

Previous successful trials of a similar system showed that it is also possible to count the number of pedestrians waiting at key junctions. The system (referred to as Ped-SCOOT) is made possible by a stereoscopic camera, which, with two lenses set slightly apart, simulates human vision. Look straight ahead for a moment. Cover one eye, then the other, and you'll see that each has a slightly different view of the room. Adding those views together gives you the ability to see in 3D, to determine how far away objects are, and, if you're a camera stuck on a pedestrian crossing, helps you to accurately count people. When this camera detects high numbers of pedestrians, the traffic lights will automatically update to give them an extra few seconds to cross. In addition, if no-one is waiting to cross, or if someone presses the request button and then crosses without waiting (not that this EVER happens, right?), the system detects it and resets itself. Ped-SCOOT is now being rolled out across London, with many other cities expected to follow suit. The only disadvantage to this tech is that there's now no straight answer to the question I've been asked a hundred times: **does the request button on a crossing actually do anything?**[*]

Other cities give different priorities to cyclists and pedestrians. According to the BBC, up to 70 per cent of all journeys in Amsterdam are now made by bike, thanks to huge investments in cycle-only roads. Lots of start-ups are designing technologies for the growing cyclist population, from a helmet with a heads-up display, to a lamp that projects an image of a bike a few metres ahead of the wearer – a kind of early notification for motorists. For pedestrians too, science and engineering are right at the heart of tomorrow's cities. MIT's AgeLab designed a suit that can be worn by researchers to mimic the 'physical challenges associated with ageing'. With a neck brace, vision-limiting goggles and carefully placed weights, this suit is being used to design better roads and pavements for our ageing population.

Gazing further ahead is risky. There are many who are convinced that every single vehicle on our roads will be fully autonomous within the next 20 years. I shall retain my scepticism on that one. As we've already touched on, and will cover lots more in the next chapter, driverless cars are coming, but for the foreseeable future, they'll be sharing the road with (increasingly connected) cars driven by humans. This will be a challenge for all concerned – if a human driver approaches a junction with damaged signals, they can edge forward cautiously. A computer-driven car cannot be programmed to do this (yet, anyway). The general consensus is that fully

---

[*] If Ped-SCOOT is installed on a crossing, you don't need the button at all. If not, one of two things happens. At peak hours, most cities use fixed crossing-times for pedestrians (i.e. the button is useless). At off-peak times, unless you push the button, the pedestrian green light will not appear. In summary: if there is a button, just push it.

autonomous vehicles should be safer for cyclists and pedestrians, but as yet, no large-scale real-world trials have been carried out.

In the midst of this chaos, data will be the thing that bridges the gap between the near-term and the more distant future. For traffic monitoring and management especially, the next step will be to develop systems that can 'learn' from their environment – where measured data can be put back into the computer model to make it ever more accurate. To find out more about machine learning, I spoke to Dr Simon Box from the University of Southampton. He is developing a traffic control system that learns from humans. You may have the impression that the Age of the Human is practically over, that computers can outperform us in almost everything and that our days are numbered. According to Box, this is simply not true. 'Some problems would take any computer an impossible amount of time to solve. Traffic control is one of them. The best we can do is find a compromise, an approximation to find as good an answer as possible. And this is something humans are rather good at.'

Box designed a computer-game-like interface that allowed people to 'play' with a simulated road network. Players watch individual vehicles driving through the light-controlled junctions, and after some time, are encouraged to make changes to the switching routine, to improve the flow of traffic. By capturing the player's strategies, the computer learns what works and what doesn't. Successful strategies can then be fed back into the traffic optimisation program, meaning that it improves each time. While promising, the model has some limitations – so far, it has been demonstrated only on a small network containing three junctions. But Box has been discussing its use with the team at TfL, so perhaps we're close to seeing traffic lights that truly learn... with a little help from humans, of course.

This is still a relatively short-term vision. If we take driverless cars to their logical endpoint, it's clear that the road network will need a complete redesign. With no 'drivers' to inform, what use is a traffic light, stop sign or box junction? With sensors embedded everywhere, the road's main job will be to provide a smooth, robust surface to travel on, while acting as a communication platform for our vehicles. Much more on this in the very next chapter.

Pavement

For too long, pavements have just sat around, nothing more than a passive surface for us to stroll lazily along (or silently rage along, when stuck behind lazy strollers). But in the future, we'll make them work for us – they will produce electricity! Right now, piezoelectric materials are being trialled for use in both paving slabs and rubber floor tiles in train stations.* These materials produce small amounts of electricity when they are vibrated, squeezed or stretched. Commuters in two train stations in Tokyo are already testing them out, and while there's no doubt that it's a cool idea, it's time to don your sceptic's cap to discuss an important caveat.

The thing about energy is that you can't get something for nothing. If you're generating electricity, the energy for it must have come from somewhere. In this case, the energy you scavenge comes from people's footsteps, and the energy people use while walking (kinetic energy) comes from the food they eat. Now, in order to produce electricity from a piezoelectric material, you need it to bend slightly, so a footstep-harvesting slab will feel slightly bouncy or squishy underfoot. Think about the last time you had to run along a soft or sandy surface. If you find it more tiring than walking on a hard surface, it's because you're having some of your kinetic energy scavenged!

There are a number of companies who already produce energy harvesting paving slabs; some are based solely on piezoelectric materials, and others on piezos plus an electric generator. All the companies are extremely cagey about how their systems work, which I find a bit frustrating. Various publicity quotes suggest that a single footstep, which

---

* Piezoelectric materials get their name from *piezo* (Greek), which means to press, and *electrum* (Latin).

compresses the tile by 5mm (0.2in), could produce up to 8W of electricity. A quick calculation shows that this number may be a little generous... but let's just go with it for now. To get even close to 8W, the slab would need to be compressed (stepped on) twice a second. This is plausible at peak times in some of the world's busiest train stations, but probably not outside of those hours. And what can you do with that much power? Well, it's roughly equivalent to the power consumption of a single nightlight, which I think we can all agree isn't a lot.

To capture any useful amount of energy, you'd need to have these compressible tiles covering vast areas of the ground – be it on our streets or on our station concourse – and right now, their cost is too high to even consider that. I also have a sneaking suspicion that commuters might start to expect a discount on their tickets, in return for losing some of their energy. Don't get me wrong, I absolutely love the concept, and I have no doubts that we'll see many of these systems in future, but maybe don't believe *all* the hype.

Air

Calgary has long been viewed as the centre of Canada's oil industry. But in the midst of this land of fossil fuels, a young company called Carbon Engineering are looking at a different future – one that involves cleaning the air of carbon dioxide (CO_2). Despite its relative rarity in our atmosphere (it makes up just 0.04 per cent of it), CO_2 knows how to hit headlines, especially for its role in climate change.[*] But carbon dioxide is also vital for life, as you might remember from school biology – plants and algae need it to make food, and happily, they produce oxygen as a by-product. This part of the Earth's natural carbon cycle has, for around 3.2 billion years, regulated the exchange of carbon through our atmosphere, oceans and ecosystems. CO_2's presence in the atmosphere is what helps to trap the sun's light and keep us

---

[*] For every CO_2 molecule you find in the atmosphere, there are almost 2,000 nitrogen molecules.

warm. However, as has been well documented, since humans entered the Industrial Age, levels of carbon dioxide in the atmosphere have reached record-shattering levels, unbalancing the cycle.[*]

For Carbon Engineering (CE), one potential way to manage this is to extract the CO_2 directly from the air, as a tree does. At the front of their prototype system, a wall of fans sucks air through corrugated sheets of plastic coated in potassium hydroxide. This compound combines with the carbon dioxide in the air to produce water and another carbon-rich liquid. Several chemical steps later, CO_2 is produced in gas form, with the rest of the air released. So why is this relevant to our roads? As Professor David Keith, President of CE, said, '[This] could be one of the only feasible ways to capture CO_2 from small, mobile sources like cars, trucks and planes. Together they make up 60 per cent of today's total carbon emissions.'

The designer of CE's air capture system, Geoff Holmes, told me that 'really, our aim is to go full-scale with this. We envisage a plant that could capture 1,000,000 tonnes of carbon dioxide a year.' Given that, according to the US Environmental Protection Agency, an average passenger vehicle pumps out 4.7 metric tonnes of CO_2, this future plant could capture the equivalent of the emissions produced by approximately 213,000 cars.[†] As we discussed in Chapter 2, many of the challenges facing future cities involve 'joining the dots' to produce something that works together, as an ecosystem. So while CE's technology is undoubtedly impressive, it's important to remember that it will only ever be one part of a solution to the complex problem of climate change.

---

[*] Short summary – we are all made of carbon. When we die, that carbon is trapped in our remains – if you heat up fossils, squash them and fast forward by a few millennia, you're left with CO_2-rich fossil fuels. Burning them releases that gas into the atmosphere.

[†] CE's own figures are a little higher – they say that the average car or truck emits 3 tonnes of CO_2. If we go with that, it'd mean that their system could negate the emissions from > 300,000 cars.

But for now, I think that's enough peering into the future. We've travelled the length of many urban roads and scaled the heights of iconic bridges. And we've met some of those people whose research will transform them all tomorrow. It is time to grab your driving gloves.

CHAPTER FIVE

Drive

What good is a fantastic road network if you have nothing to drive on it? In an alternate reality, I'm swanning around in a racing green Aston Martin, but perhaps you're more of a Cadillac, Rolls-Royce or Ferrari fan. Regardless of the brand you dream of, all cars share the same basic structure. In fact, they also have a lot in common with every bus, truck and van that drives along our city streets. OK, an exhaust pipe from a 10-tonne truck won't be much use in your compact hatchback, but that's only because of its size. Fundamentally, every road vehicle exists to move people or goods around, so while I might take the (charmingly) lazy option of just referring to 'the car', I'm really talking about everything with tyres and an engine.

With that in mind, let's analyse the car's DNA, to understand what it is that makes them go *vroooom*. As usual, we'll have lots of experts on hand to guide us, because, while they look (largely) the same as they did 30 years ago, under their skin cars have been quietly revolutionised… and that's nothing compared to what's ahead. Of course, the main reason we need to talk about cars is because they and their brethren have had a major role to play in shaping the world's cities.

Today

Car culture really began at the end of World War II, when the US manufacturing industry moved away from munitions and created the first truly affordable, mass-produced cars. By the end of the 1950s, the number of cars in the US had more than doubled, and one in six Americans were employed, directly or indirectly, in the automotive industry. In the same decade, Britain became the world's biggest exporter of cars, and the development of post-war road networks sealed the car's place in the heart of the city-dweller.

With mass transport becoming increasingly popular in recent years, many commentators have suggested that cities have outgrown the car. Your own view on that may well depend on where you live. Beijing now has more than five million private passenger cars, one for every four of the city's residents. In comparison, over half of London households have at least one car. And what about America, the country most synonymous with the road trip? At first glance, the statistics seem to bolster its reputation – across the country, a staggering 92 per cent of households have a car. But this trend is not reflected in all of its urban centres; New York's car ownership figures are rather similar to London's. For other vehicles, the stats are just as variable. In the US, the population of trucks (broadly defined as anything with six tyres or more) has stayed pretty constant since 2007, at around eight million. As of March 2014, there were over 9,300 buses on the roads of London – a figure that has been steadily increasing over the past five years. In Beijing, bus populations were closer to 19,000, but that hasn't really changed since 2008. However you feel about getting more cars off the road, it seems that buses and trucks aren't going anywhere.

In many cities, it's not mass transport that's driven people away from private car ownership – it's the availability of group schemes. In cities all over the US, Canada, the UK and mainland Europe, centrally managed 'car clubs' are growing at an unprecedented pace. Less like a rental car and more like a communal car for the neighbourhood, you can book one only when you need it. However, those who do choose to buy their cars are keeping them for longer – at the end of 2013, the average age of vehicles on America's roads was 11.4 years, three years older than it was in 1995.

We also need to broach the difficult subject of weight/mass gain. Between the 1950s and the 1970s, the average American car became streamlined, consistently losing weight year on year. But in the three decades that followed, the trend slowed – cheap petrol (gasoline) made cars less expensive to run, and in response, manufacturers started adding stuff back in. This means that despite all of our engineering progress, today's average car now weighs almost exactly the same as it did in 1975, approximately 1.8 tonnes. Our love of bigger, more comfortable vehicles has a lot to do with this, as do the added components needed to meet today's safety regulations. However, alongside the increase in weight, manufacturers managed to hugely improve the amount of power that can be delivered by a car engine, while still (and this is what really defines the modern car) improving fuel economy. In short, today's car may be heavy, but it can do a whole lot more with that weight – it can accelerate faster, get more from its fuel, and produce lower CO_2 emissions.

Of course, these days, we see a far greater diversity of vehicles on the road than those running on fossil fuels. Hybrid and electric vehicles are gaining in popularity, and according to the European Commission, biofuels represent around 5 per cent of all transport fuel used in the region. Germany, Italy and Denmark are looking even further ahead – all three countries have a significant number of hydrogen refuelling stations for those vehicles that run on fuel cells. With all of these technologies now on our roads, and many of them expected to become the norm in the future, I think it's about time we figured out how they work.

Fuel

Petrol (gasoline) and diesel vehicles are by far the most common. Between them, they represented 98.5 per cent of the UK's car market in 2012, so they're probably a good place to start. In simple terms, a petrol or diesel engine converts the chemical energy in the fuel into kinetic (or motion) energy. It does this using a process called internal combustion, which was originally patented in 1861 by French inventor Alphonse Eugène Beau de Rochas (who may have the best name ever). Internal combustion involves injecting a small amount of a high-energy fuel into an enclosed space where it is ignited, releasing a huge amount of energy that can power a car.

That top-level description might be enough for some, but I know that you, my lovely readers, will want more, so here are the details. In order for an internal combustion system to work, we first need a combustible fuel – one that is able to ignite safely. Both petrol and diesel fit that description, which is unsurprising given their similar chemical make-up. Both are hydrocarbons, meaning that they consist entirely of hydrogen and carbon atoms, arranged into long chains or complex shapes. Scientists came around to hydrocarbons by a process of elimination; the inventor of diesel, Rudolf Diesel, originally wanted to use coal dust as fuel in his compression-ignition engine. He also showed his early futurist credentials by experimenting with vegetable oils – in fact, peanut oil was used in the engines he exhibited at the 1900 World's Fair in Paris.

It was German engineer Nikolaus Otto who really made combustion a success. He realised that if the fuel is compressed before it is ignited, the release of energy can be even more spectacular.[*] How fuels respond to this treatment depends on its octane rating, a measure of its performance. But to understand what this actually means, let's talk about the reality of a working engine. Take the four-stroke engine – so named because it uses a four-step process. In a petrol engine, these steps are most easily remembered as suck, squeeze, bang and blow (stop giggling at the back please!).

- **Suck**: First, air is taken into an engine cylinder, and a small amount of fuel is added.
- **Squeeze**: This air-fuel mixture is then squished into a smaller volume using a moving arm called a piston.
- **Bang**: Then, a component called a spark plug produces a spark, which causes the mixture to ignite.
- **Blow**: This explosion forces the piston back down. When it hits the bottom of the cylinder, the air intake again opens and the spent gas escapes through the exhaust valve… and the cycle repeats.

---

[*] Too spectacular sometimes! Many of the earliest engine pioneers were almost killed by exploding fuels.

In order to get as much power as possible from your fuel and to create a practical engine, you need a system with hundreds of these miniature explosions happening every minute. This is why all modern cars have more than one cylinder – for example, a V8 is a car with eight cylinders. So what role does a fuel's octane rating play in all of this? Contrary to what you may have heard elsewhere, it has nothing to do with the energy content of the fuel. It's all about how much you have to squeeze it before it combusts.

Fuels with a high octane rating tend to be used in petrol engines, while those with a low octane rating are used in diesel engines. This is down to an important difference between how the engines operate. Instead of compressing a fuel-air mixture, diesel engines only squeeze air (heating it up). The fuel is injected afterwards, which makes squeeze-free, low octane fuels the perfect choice. Because it's the heat of the compressed air that lights the fuel, diesel engines don't have spark plugs either. It's best not to mix them up – using a low octane fuel in a petrol engine causes 'knocking' (a pinging sound resulting from the fuel igniting earlier than planned), which messes up the whole cycle, and could eventually cause engine failure. And no, using a high octane fuel in a diesel engine will not 'boost its performance'. The engine performs best with the fuel it was designed for. Don't believe anything that says otherwise.

Regardless of the fuel used, by attaching a crankshaft – a long metal rod – to these cylinders, the piston's up-and-down, or linear motion, can be transformed into the rotational motion needed to turn the wheels of a car. The clutch allows you to control how much of the engine's power makes its way to the wheels. It's made from three different components that together act like a friction switch between the engine and the crankshaft. When you put your foot on the clutch pedal, you separate the components from each other – this lets you safely ramp up or down the engine, while the wheels keep moving. Releasing the pedal brings everything back together again to move as one unit (only this time in a different gear). Cars combine all of these steps to turn fuel into motion. But, even the very best engines cannot convert all of the fuel's chemical energy. Typically they manage only a small fraction, with most of the rest lost as heat energy. And while this might surprise, shock or horrify you, we'll soon uncover ways to take advantage of that wasted heat.

One vital component of a diesel or petrol car is its catalytic converter, which transforms some of the engine's harmful gases into less harmful emissions. Catalytic converters look rather like a dense white sponge, coated in a thin layer of platinum or palladium (the catalyst). As soon as a car starts running, the fuel begins to break down into various gases that travel along the exhaust pipe, until they meet the catalytic sponge. There, they do a bit of chemistry – nitrogen oxides (with the formula NO_x), long linked to lung problems, smog and acid rain, are turned into nitrogen and oxygen. And carbon monoxide (CO), which if inhaled can cause suffocation, is converted into carbon dioxide (CO_2)... OK, so that one isn't quite so clean in the end, but it's still an important improvement. The tech hasn't changed much since the 1970s though, and with the cost of a gram of platinum around \$28 (£20) at the time of writing, catalytic converters are due a reboot.

Plug

Hybrid and fully electric cars (EVs) are an increasingly popular sight on our urban streets, with many major cities investing in public-access charging hubs. As the name might suggest, *fully electric* vehicles don't have a combustion engine at all; they run on three main components – batteries, which can be charged from the mains, a controller (basically a box of electronics) and an electric motor that physically turns the wheels. On all cars, the driver can interact with the engine via the accelerator pedal – if you want to speed up, you 'put your foot down' to let more fuel enter the cylinders. But in an electric vehicle, that pedal is connected to the controller – the brains of the operation. The controller manages the voltages from the battery, reads all of the data from the pedal, and continually calculates and adjusts the electrical signal, so that the motor gets what it needs to turn the wheels. In short, an electric vehicle has very little to do with plumbing, and a lot to do with wiring.

Diesel and petrol cars use batteries to start up, and to power various components, but for electric vehicles (EVs) *they are the fuel* – if they fail, you're going nowhere. So the design of the vehicle's battery pack is a major consideration in EVs. Before we go into that, let's take a small step back to remind ourselves how batteries work. It's really all about electrons. Each battery 'cell' has three main ingredients: a negative electrode (the cathode), a positive electrode (the anode), and the electrically conductive liquidy-stuff that separates them (the electrolyte). The cathode is negatively charged because it has too many electrons, while the anode has too few. When both sides of the battery are connected to a device, electrons flow from the cathode, through the electrolyte, to the anode. This flow of electrons is what we call current electricity, as you'll hopefully remember from Chapter 2. The battery's performance and lifetime depend on the materials used for the electrodes and the electrolyte. The cheap AAA batteries lying in your drawer are likely to have electrodes made from zinc and carbon. The fancier and considerably more expensive batteries in your laptop are probably based on lithium.

No matter how fancy they are, all batteries eventually run out of charge, and you really don't want this to happen when your electric car is in the middle of nowhere. As a result, battery capacity is still the major limiting factor for electric vehicles. EV batteries are also very expensive – in many cases, a third of the total price of the electric car itself. In terms of performance, top of the electric vehicle pile is the Tesla S, the 'world's first premium electric sedan', developed by Tesla Motors.* The Tesla S uses the aforementioned lithium–ion batteries to do its thing. While a typical laptop will use four or six lithium cells, each Tesla S contains 6,831! At the time of

---

* Tesla Motors is named after Nikola Tesla from Chapter 2. And yes, I still love him.

writing, their best battery pack could manage 435km (270 miles) on a single charge, which is pretty damn impressive. However, as with all things marketing, we need to take these figures with a pinch of salt – they assume perfect driving, smooth roads, no traffic and no wind. In reality, the range is likely to fall somewhat short of that. But it remains the best in class.[*]

Most electric vehicles also employ regenerative braking to recharge the batteries on the go. In a standard car, every time you step on your brakes, your car's kinetic energy is transformed into friction by the brake pads, and it's this that brings you to a halt. But in a regenerative braking system, when the driver steps on the brake, it triggers the motor to run backwards, slowing the car's wheels. Thanks to electro-magnetic induction, an electric motor working in reverse acts like a generator. So your brakes produce electricity that can be fed back to the vehicle's batteries.

While great strides have been made to improve the range of these batteries, and reduce their cost, we're still some way from making fully electric vehicles our first port of call for low emissions. These days, that plaudit goes to the hybrid car, which takes the internal combustion engine from a fuel car and the electric motor from an EV, and slams them together. The most famous manufacturer of hybrids is Toyota, and in July 2015, they sold their eight-millionth hybrid vehicle. Their Prius model is what's called a parallel hybrid, which means that they can run on either a fuel engine or an electric motor. The other option is a series hybrid – it runs off the batteries all the time. The small fuel engine on-board just spins an electric generator.

So which one's better? Series hybrids are simpler to run and their engines are efficient, but their need for many batteries makes them expensive. Parallel hybrids make a lot of sense if you drive equally in the urban jungle and on the open road. They can run on battery power when zero emissions are required, as in the city, and switch to the engine once you need

---

[*] In April 2016, Tesla announced the launch of their Model 3. The battery range is likely to be higher than that of the Tesla 5, but details hadn't been released by the time of writing.

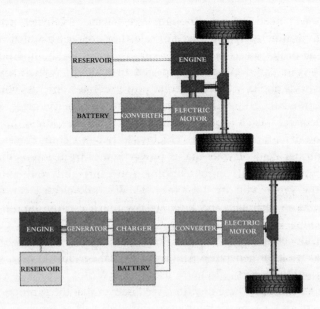

Figure 5.1 This image shows a parallel hybrid on the top, and a series hybrid on the bottom.

to reach higher speeds. Their smaller battery pack reduces cost, but it also means they're more dependent on charging. Of course, hybrid systems aren't just for cars – as of mid-2016, 1,700 hybrid double-decker buses roamed the roads of London, making up a fifth of the total bus fleet. They're split evenly between series and parallel, and Transport for London (TfL) tell me that, so far, the performance of these buses is roughly comparable. However, it is possible to design a hybrid engine that combines the performance of both series and parallel engines. This 'hybrid-hybrid' system acts like a series hybrid at lower speeds (for increased efficiency) and like a parallel hybrid at higher speeds (for increased power). While still in early development, this combination has the potential to outperform either of the existing systems. Outside of mass transport, hybrids are making their mark too – delivery giants FedEx are rolling out hybrid vans across their global network, and more than nine heavy-duty vehicle manufacturers now offer hybrid systems.

We need to ask though, are electric vehicles truly environmentally friendly? I'm told they are, again and again. The only way to find out for sure is by asking the question, 'If your car needs to be plugged in at night (which some hybrids don't), where does that power come from?' As you'll remember from Chapter 2, no two electricity grids are identical – each is powered by many different sources. So, if you drive a fully electric vehicle, and live in a country where 100 per cent of your mains power comes from low-carbon resources, then congratulations! You can rightly talk about how 'green' you are. Let's take BMW's fully electric car, the i3, as an example, and look at how much carbon dioxide it generates in charging its batteries. The UK national grid produces the equivalent of 500g of CO_2 for every kWh of electricity it generates. And from the car's technical data, we know that every kilometre uses 0.129kWh of battery power. If we multiply these together, we discover that the i3 produces 64.5g of CO_2 every kilometre – around half as much as the average petrol or diesel car in 2015.

However, this is only true for the UK grid, which is powered by a broad mix of technologies, including fossil fuels and renewables. In countries like India and China, where the grid is entirely coal based, running an electric car could even produce *more* CO_2 than a petrol or diesel car. This is true for much of mainland Europe too, according to EV expert Senan McGrath. In our chat, he told me that if we all switched to electric vehicles, but continued to generate electricity as we do now, 'there would be no more than a very marginal saving in emissions'. Lots of calculations also leave out an oft-forgotten aspect of an electric vehicle's environmental-friendliness: its manufacture and production. It's extremely likely that the materials that make up the car's components will have been produced, shipped and assembled in plants powered by fossil fuels. The rare elements, such as the lithium, will have been extracted from the ground using heavy plant machinery. There's really no straightforward way to include this in our maths, or to fairly compare these costs with those of a conventional car, but it's worth remembering that even the greenest tech can have a murky past.

What this means is that the biggest challenge facing electric cars is the grid that they plug into. Right now, we can't say that they're the right choice for everywhere. As we move away from fossil-fuel power plants and towards lower-carbon options like hydroelectric, solar and wind power, this will undeniably change. And frankly it can't come soon enough.

Bio

According to Wikipedia, a biofuel is 'fuel that contains energy from geologically recent carbon fixation'. This definition is a bit vague, so let's unpick it. Living organisms like plants are pretty awesome at taking in carbon (in the form of CO_2) and converting it into complex organic compounds like hydrocarbons. This is the process we call carbon fixation, and its job is to provide the fuel that plants need to live, grow and develop. When they eventually die, this fuel can be collected and used elsewhere. This brings us on to 'geologically recent'. Fossil fuels, such as oil and gas, are the ancient remains of once-living, carbon-fixing organisms, which over millions of years have been squeezed and heated under the Earth's surface. In contrast, biofuels are young. Really young. As in, they've only just been harvested, mid-carbon fixing. So, we could say that biofuels are still-alive, un-fossil-like fossil fuels. Because we don't have to wait around for the ingredients to 'cook' in the Earth's pressure cooker, biofuels are much easier to get hold of than fossil fuels too. But does that mean they're any better for the environment?

The debate around biofuels rages on, but before we go there (and oh yes, we will go there), we need to talk about their current status. Brazil is the world's largest producer of ethanol fuel from sugar cane, and at the industry's peak in 2010, Brazilian motorists used 22 billion litres (4.8 billion gallons) of this bioethanol. Producing it from sugar cane is relatively straightforward – once harvested, a sugar-laden liquid is extracted from the cane, and it's filtered and fermented to produce an alcohol (ethanol, in this case). Ethanol doesn't contain as much chemical energy as petrol (gasoline), so it tends to be blended in with other fuels. It's been used this way for a while – during World War II when petrol was scarce, small amounts of ethanol were added to make the precious fuel go further without greatly affecting its performance. Today, there are *no* light transport vehicles (*i.e.* cars) in the country that run entirely on fossil fuels. Yes, you read that right, *all* Brazilian cars are 'biofuelled' to some extent, with the most common blend being E85 (85 per cent ethanol).

But if my earlier rant on octane rating taught you anything, it's that you can't just whack any old fuel into your car. Most of Brazil's cars are flex-fuel vehicles (FFVs) and they can cope with a large range of fuel blends. Their engines are fairly similar to a typical petrol-only model, but they have a bigger fuel tank (lower energy content per litre means you need more fuel) and their suck-squeeze-bang-blow cycle is a little different. From the outside, FFVs look and sound like petrol-powered cars, which has made them very popular in Brazil. Government legislation and artificially cheap oil in 2011–14 led to a general downturn in the demand for bioethanol, but as of 2015, it appears to have regained its priority status. It is now mandatory in Brazil for a vehicle to run on at least 27.5 per cent ethanol.

Another widely cited biofuel is biodiesel, which is made from plant or vegetable oils – everything from sunflowers to soybean. The major benefit of biodiesel over ethanol is that it packs more of a punch – its energy per volume is pretty close to that of regular diesel. Like ethanol fuel, biodiesel can be produced in different grades, with 100 per cent biodiesel referred to as B100, and a blend containing 2 per cent biodiesel labelled as B2. Generally, blends that contain 20 per cent biodiesel and lower can be used in standard diesel engines with little or no modification, so these tend to be the ones we see most.

Now that we've met a couple of biofuels, let's consider how 'green' they really are. There's a common misconception that the words 'renewable' and 'green' are interchangeable, but they're not (please think of me when you say that to someone over dinner). Biofuels are definitely renewable – plants can be replanted – but burning them produces greenhouse gases, so they can't truly be considered 'green'. Long before biofuels make their way into an engine, their raw ingredients need to be planted, nurtured and harvested. This all requires energy, which will come at a cost to the environment. And then there's what comes out of the exhaust pipe. It's claimed that biofuel vehicles emit fewer harmful gases than those run on petrol. As with most things, it's just not that simple. E85 definitely produces less carbon monoxide than a petrol car, but it produces more acetaldehyde (highlighted as a potential carcinogen) and ozone. In that context, how do we define the better option?

It is when we get to the land required for biofuels that the debate gets very one-sided indeed. Currently, most biofuels are produced from crops that are also food crops (such as corn, wheat, sugar cane, sugar beet, palm oil, rapeseed, soy). This is definitely not sustainable. As we'll talk about in Chapter 7, the world has a finite supply of fertile, farmable land, which is already under immense pressure to produce food. Handing over a large proportion of it to grow vehicle fuel is simply not an option. Even in Brazil, land of the biofuel, huge political debate rages on the subject, with arable land and sections of rainforest at risk. In 2014, the European Commission announced that one-fifth of all of Europe's electricity would come from renewable sources by 2030. This immediately pushed biofuels up the political agenda, but research from the University of Vienna showed the realities. If we were to use the grow-for-fuel approach, the total land footprint needed to hit that target would be 70.2Mha (271,000 square miles) – an area the size of Poland and Sweden combined. Can Europe's stomachs forgo that much food in return for biofuel? Short answer: nope.

Thankfully, there are alternatives. Across the world, pilot projects are looking into using waste as a source of biofuels. My fellow Irish science nerd, Finn Coyle from TfL, told me that they are 'specifically looking into the use of animal tallow and used cooking oil, which are both by-products from the food industry, as potential fuels for our buses'. Coyle and his team are not alone either – ethanol made from corn waste is being touted as a real alternative to petrol in the coming decade. More on that in a little while, but first, the most futuristic fuel of all.

Hydrogen

Hydrogen is the lightest element in our universe: it's as small as it gets when it comes to atoms. And because it is very reactive, it's all around us, locked up in water, organic matter (such as us!) and in those hydrocarbon fuels we talked about. But hydrogen is far more than just an abundant, tiny, reactive atom – as we discovered in Chapter 2, it could have a huge role in our future energy landscape and might also bring about the biggest change in cars we've seen yet.

Fuel cells are often described as 'hydrogen batteries', but unlike a typical battery that contains a fixed amount of charge, fuel cells need a flow of both hydrogen (H_2) and oxygen (O_2) gases to work. Inside the cell, electrons are stripped from hydrogen, leaving behind a positively charged particle (H^+). The newly freed electrons flow out of the cell, producing current electricity that can be used to power the car's motor. The positively charged leftovers are combined with oxygen gas to produce water (H_2O); most of this is recycled in the cell, while some is lost as water vapour through the exhaust.* Fuel cells do chemistry 'on the fly', continuously stripping electrons from the billions of hydrogen atoms in the gas, and producing water as a by-product. No CO_2, no NO_x, just water. You can probably begin to see why car manufacturers and governments are prioritising their development.

I wanted to learn more about automotive fuel cell technology, and to understand the challenges still facing them. Luckily, one of my best buddies just happens to be a world-leading expert on everything to do with fuel cells. Dr Gareth Hinds from the National Physical Laboratory

---

* If all of this sounds vaguely familiar, it's because fuel cells are mini-electrolysers (from Chapter 2) operating in reverse! Instead of using electricity to split water into hydrogen and oxygen gas, fuel cells combine hydrogen and oxygen to release electricity.

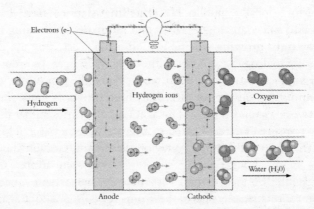

Electrons (e-)

Hydrogen   Hydrogen ions   Oxygen

Water (H₂0)

Anode   Cathode

Figure 5.2 In a fuel cell, hydrogen and oxygen are combined to produce electricity and water.

(NPL) agreed to answer all of my questions, so we headed to the place where numerous scientific discussions have occurred – the pub. The first thing that I learned was that, in theory at least, a fuel cell car could travel much further between refills than your average electrical car: commercially available systems offer up to 500km (310 miles) on a full 'tank' (again, assuming perfect driving conditions) and take little more than five minutes to refuel.

The second surprise came when we began to talk about the status of various car manufacturers. According to Gareth, many have now matured their fuel-cell technology so much that 'we'll soon see their first production models on the road on a regular basis'. Perhaps even more surprisingly, major manufacturers are openly collaborating on fuel cells. The HyFive project (tenuous acronym alert! It stands for **Hy**drogen **f**or **I**nnovative **Ve**hicles) includes car giants Hyundai, Toyota, BMW, Honda and Daimler. They're working with research institutes to develop the cells, and with European governments to help establish the necessary refuelling infrastructure. Without hydrogen fuelling centres, fuel cells won't find widespread use, so this project seems to tick both boxes. And excitingly, one of the first refuelling stations is at NPL, the lab at which I spent many happy years as a scientist.

More locally, public H_2 refuelling stations have been around since late 2013 – Japan and Germany are leading the way, and a growing scheme in California is catching up. But storing hydrogen gas is not without its challenges. In order to fill your on-board hydrogen tank as quickly as possible, the gas must be delivered at incredibly high pressures – anywhere between 350 and 700Bar. This is at least 100 times higher than the pressure of the gas stored in a typical aerosol can... Think of that the next time you spray on deodorant. Compressing a gas is all about fighting the forces between atoms. Gas molecules bounce around very happily, but as you squeeze them into a smaller space, the atoms try to spread out again. The closer the atoms are to each other, the harder they push apart. For H_2 gas, this means we need a storage tank with very strong walls. There is no need to panic though – we've been using high-pressure gases everywhere for decades, so we understand their complexities. And trust me, the safety regulations involved in installing gas refuelling centres are very thorough!

Anyway, assuming that we can establish a global network of hydrogen fuelling stations, where will the fuel actually come from? Well, the oxygen bit is easy – fuel cells can just use air. However, hydrogen has to be extracted from other compounds, which is much trickier. Right now, the most common way to produce hydrogen is steam reforming, where super-heated steam is reacted with methane. This approach is ridiculously costly in terms of both money and energy, and could never be described as clean (methane is a greenhouse gas). But, as we discussed back in Chapter 2, electrolysers, powered by the excess electricity from wind turbines, can extract hydrogen directly from water. These efforts make a lot more sense, and thankfully, they're picking up pace. So now we have the fuel and the means to deliver it. The only thing left are the cells themselves.

Most of those developed for cars are called polymer electrolyte membrane (PEM) fuel cells. A single PEM fuel cell is like a sandwich: the slices of bread are two platinum-based electrodes and the filling is a solid polymer electrolyte. Each produces a very small amount of electricity, so you need stacks of them to get enough voltage to run the motor, which is what moves the car. Current PEM fuel cells are pretty efficient, relatively simple but very expensive. As I type, universities on every continent are researching alternative materials, so this is very much a work in progress. And of course, unlike combustion engines, they don't produce any polluting hydrocarbons, which is why fuel cell buses are described as having 'zero tail pipe emissions'. Wanting to learn a little more about them, I spoke to TfL's Director of Bus Engineering, Gary Filbey. As of late 2015, London had a fleet of 10 hydrogen fuel buses running on route RV1, and their performance has improved throughout the pilot. 'Our fuel cell buses used to be able to run for about eight hours a day,' said Filbey, 'but now they can run a full duty cycle (anything up to 18 hours) and take just a few minutes to refuel.' TfL has its own maintenance facility and H_2 refuelling point within a huge bus depot in north-east London. Every night, their fuel cell buses go there to have their tanks refilled. London is not the only city looking into the practicalities of fuel cell buses. As of early 2015, there were 22 active demonstration projects in cities across the globe.

Road vehicles have shaped us, and our cities, for almost a hundred years. So, now it's time to consider what the future holds for them.

Tomorrow

Well, obviously, within the next 20 years, we'll be able to 'drive' our flying car to the office and pack it up in a briefcase when we arrive (like the Jetsons). Or not. Sorry to disappoint, but wheel-based cars, vans, buses and trucks will remain a huge part of the urban landscape in the coming decades. While many city-dwellers live very happily without cars, they might struggle without a bus service, and as an avid internet shopper, I welcome delivery vans with open arms. These vehicles will continue to move stuff and people around cities. While they'll mostly look the same on the outside, under the bonnet (or hood, whichever you prefer) they'll be much sexier than today's fuel-belchers.

Undoubtedly, the most dramatic change will come with autonomy in cars. In the short-term, they'll just get smarter – safer to drive, more efficient, lighter, and composed of long-life recyclable components. Alongside this, we'll see better, more integrated systems being developed across the global road network. Eventually though, cars seem to be destined to become fully-driverless.

Body

Let's start with those changes that we'll see very soon – they're all about materials. In 2012, the US government introduced new regulations on fuel efficiency, and in 2013, the EU followed suit. While the numbers in the regulations differ, both say that by 2025, all new cars should be able to travel further using less fuel. The US is targeting 54.5 miles per gallon. As of 2015, the average fuel efficiency for new cars stood at about 25mpg. You don't need me to do the maths on this one – it's a challenging target that even some hybrids fall short of.

Unfortunately, we can't look solely to the engine to fix this. We could also improve efficiency by reversing the trend we discussed right at the start of the chapter – we need to start making vehicles lighter again. A lighter car can run on less fuel, and given that the outer shell (or body) of a car accounts for about one-third of its weight, it would be a good place to begin lightweighting. Many cars use bulk steel, making them strong but rather heavy. Recent research in Germany used lasers to harden sections of thin, lightweight steel sheets. This made the sheets twice as likely to survive an impact as the standard steel sheets used today. A more future-facing (and even lighter weight) option is aluminium – Aston Martin, Land Rover, Ferrari, Ford and Audi have all launched aluminium-body cars in recent years. This one change has resulted in considerable weight savings; 315kg (695lb) in the case of the Ford F-150 truck. But while it's popular for use in the bonnet (or hood, whichever you prefer), aluminium can't be used everywhere. It may be lighter than steel, but the downside is that it doesn't fare as well in a crash. Cue another bit of material science: **Young's modulus**, or **E** (named after the nineteenth-century scientist Thomas Young). It tells you how much a material yields under the application of a force. Rubber bands have very low E because you can bend, stretch or squeeze them with little effort. Metals tend to have much higher E because they're harder to deform, but aluminium buckles under a third of the force needed to bend steel. However, by adding other metals (e.g. zinc, magnesium and copper) to aluminium, and heating the resulting mixture, you can get a super-strong, ultra-lightweight aluminium alloy that's perfect for use in cars.

Looking even further ahead, it seems that the material-of-the-moment, graphene, might play a role in cars too. A huge European project, called *iGCAuto*, is investigating the potential crash-worthiness of a graphene-based polymer material. I'm sorry if it feels like I've been hammering on about graphene – it's just that I think it is a little bit sexy (there, I've said it). As you might remember from earlier chapters, graphene is very, very thin (less than 0.0000005mm thick) and lightweight, but despite this, it is incredibly strong. As with everything nano, the major issue is how to make it practical, and right now, we're pretty far from that. Professor Ahmed Elmarakbi, lead scientist on the project, has said that 'The issue is not only producing graphene-based products, the issue is applying them on a large scale in cars.' But if they manage to do it, and produce graphene materials that can withstand a crash, it may very well change the way we build vehicles.

In late 2014, Mazda announced that they'd made a plant-based bioplastic material suitable for use on the exterior of a car. At the time of writing, things had gone a bit quiet on that front, but there's no denying it's a pretty cool idea. Many other car manufacturers are looking at them too, but the current price of bioplastics means that they're restricted to high-end models. In terms of their structural properties, it's hard to find anything on those specifically being designed for

cars. There is a growing number of research papers on
bioplastics though, and the figures look optimistic. But
growing plants to make plastics seems a bit stupid, doesn't it?
Especially in light of the argument on biofuels. Thankfully,
much of the leading research isn't looking to plants; instead,
recycled rubber, food waste and even algae are being mined
to create sustainable plastic components. This work has much
wider implications than just the car – almost all of the plastic
around us today is made using the products of the fossil-fuel
industry. If we find cleaner, greener ways to produce similar
materials, the difference will be felt in every aspect of city
living. Our lunchboxes, phone cases, streetlamps, payment
cards, keyboards and lift buttons could all be subtly but
dramatically transformed.

To build the cars of tomorrow, we'll need a new generation
of materials that are both lightweight and crash-safe. It's
likely that the ultimate solution lies in using mixed materials –
some components made with steel or aluminium, with others
made from composites and bioplastics. All of this will be
relevant no matter what's under the bonnet, since every
vehicle benefits from lightweighting. There is of course the
question of how cars and buses will look in our futuristic
cities. I'm the very last person who could provide sketches
of swishy concept cars, but if we are heading towards a
driverless-car-filled future (and more on that shortly), will
we even need steering wheels?

Power

Before we look into all of that, let's burrow under the surface
a bit, and talk about what will power tomorrow's vehicles. As
much as it pains me to say this, I really don't think we'll see
a wholesale flip to 'cleaner' vehicles within the next decade.
Yes, the electricity grids of many cities are being decarbonised
at a record rate, but initially at least, I think we'll see no more
than a slow and steady move away from fossil fuels. However,
if we are to reach the ambitious emissions targets set
for cars and other vehicles, the tortoise approach just isn't
going to cut it. Once we begin to see a considerable

decrease in the cost of batteries and fuel cells, the trickle away from hydrocarbons will become a flood. However 'easy' it is to run a car on these systems (and as we've discovered, it's still pretty tough), the challenge will be much greater for vehicles like delivery trucks. They're big and heavy and need a lot of energy to get moving. Even our very best batteries struggle at the moment, so in the short term we're likely to see clever stop-gap solutions for these beasts of the road. One example of this is a novel catalytic converter developed by Dr Ben Kingsbury. This new device could even cut fuel consumption, thanks to a clever bit of redesign. A converter with small channels is good – it means the gas spends longer flowing past the catalyst, so it gets cleaned efficiently. But if they're too small, gases build up in the channel, which forces the engine to work harder and burn more fuel.

Kingsbury's solution was to increase the size of the channels, while making them more complex. On first glance, his converter looks like a bunch of straws stuck together, but in fact, every 'straw' is more like a ceramic thorny branch, with many spikes or channels sticking out of its walls. In lab tests, the prototype used 80 per cent less of the expensive platinum catalyst, but was more efficient than standard catalytic converters. It's expected that in real operating conditions, the converter will need a little more catalyst than this, but it will still use substantially less than today's systems. When I spoke to Ben, the team were talking to car manufacturers to turn this into a commercial reality, so keep your eyes peeled for this one – it might just replace ageing systems on cars and trucks the world over.

Of course, as we know, internal combustion engines can run on stuff other than fossil fuels. Biofuels developed from wastes (such as cooking oil), agricultural residues, forestry residues and novel feedstocks, like algae, are on the march. As I mentioned back in Chapter 3, there are even a few buses that run on poo. One of my favourites is the aptly named No. 2 bus in Bristol, UK. It is powered by a methane-rich gas extracted from processed human faeces, and its refuelling point is the local sewage plant. Inside the bus, the gas is compressed into a liquid before being injected into the engine (like any other fuel). It can travel 300km (190 miles) on a full tank of biogas equivalent to the annual waste of around five people. And before you ask, no, the poo-bus doesn't smell at all! This kind of project is more than just PR – by using waste products, it shows that biofuel production doesn't have to compete with food production. It can be another way to get something of value from what we usually discard.

For the electric vehicle, the future is all about improving the reliability and range of its power source. Not only will better batteries lead to more practical cars, they might also shape the urban landscape – plugging in is *so* 2016; in-road charging is the way forward. Regardless of which method we use, a higher capacity battery will be a big help. Senan McGrath told me that 'every e-vehicle manufacturer is concerned about over-using the battery, so they use only a percentage of its full capacity'. This means that while the lithium-ion batteries of the very best electric vehicles are efficient, they're being under-utilised. And this makes the fact that they're also very expensive (and occasionally prone to serious electrical faults) all the more frustrating. Thankfully, researchers across the globe are working on this problem, and many possible solutions are being mooted. So many, that all I can do is give you a tiny flavour of them.

- A team at the Fraunhofer Institute in Germany are looking at replacing one of the lithium electrodes with one made from sulphur and carbon. Their prototype battery was lower in cost, and more stable, than standard lithium batteries. The work is still at the research phase, but the team believe that with further modifications, their battery could allow tomorrow's electric cars to travel double the distance of today's.
- Graphene also makes an appearance in future batteries. In late 2013, theorists from Rice University

showed that by adding atoms of boron to sheets of graphene, you get a material that can store twice as much charge as the most widely used electrode material in lithium batteries. As yet, no-one's managed to actually produce this material, but it will be a game-changer if they do.

- And more recently, a team from the University of California reported on a paper-like material that could be used in next-generation car batteries. Made from silicon fibres invisible to the naked eye, this material can be cycled – that is, charged and discharged – many times without being damaged. The silicon nanofibres are relatively easy to produce too, so they stand a good chance of starring in a real-life product.

If we fast-forward a little more we'll see that following in the footsteps of next-generation electric vehicles will be those that run on hydrogen fuel cells. With several car manufacturers already launching production models, the pressure is on today's governments to provide the necessary infrastructure to make it all work. As we've already seen, Germany is a leading light in this effort. They've pledged to increase the number of publicly accessible hydrogen stations in major urban centres to 400 by 2023. As we find cheaper and more renewable ways to produce high-purity hydrogen gas, the price of fuel cell cars will decrease too. Current research is looking into the optimisation of all aspects of the cell itself, from the catalyst through to the fundamental design. Together, these efforts will start to shift fuel cells away from their current niche applications and bring them into our garages. When I asked Gareth Hinds about the future challenges facing road vehicles, his response was suitably cautious: 'The only way forward will be a mix of everything – lightweighting of materials, a grid powered by sustainable sources, *etc.* ... but I would be amazed if fuel cells don't become a wider part of the transport landscape.'

Regardless of whether you plug in your car to charge it, or pull into your local biofuel or hydrogen station, in the future you'll be confident that renewable sources of energy have provided the power it runs on. Only then can electric cars and fuel cell cars be said to be truly non-polluting forms of transport.

Steal

When it comes to electricity, there is some low-hanging fruit that we shouldn't ignore. As we've talked about lots of times in *SATC*, it is possible to change energy from one form to another without upsetting the physics gods. Which means we can also capture 'waste' energy and transform it into something more useful.

Let's start with heat. Sometimes, that's the form of energy we want – when we're chilly and turn on our radiators, for example. But when it comes to cars, heat is rather low down the priority list, which is a shame when you consider how much of it an engine produces. Harvesting this heat is an area that I spent most of my years at NPL working on. And it was the following factoid that first hooked me in: in a traditional car engine, almost *two-thirds* of the chemical energy in diesel or petrol is completely lost, mostly in the form of heat energy. Have a think about how much you last spent on fuel while I reiterate that: only around a third of what you put into the fuel tank is actually used to make your car move. Much of the rest is thrown away, or at least produced in a form that we don't want. Thankfully, it's possible to transform heat directly into electricity, using thermoelectric (TE) materials.

TE materials behave in a very particular way when one side is hot and the other cool. And it's all down to how the atoms (and their electrons) jiggle under this temperature difference. At the hot end, the electrons vibrate so vigorously that they move towards the colder side, where the electrons are jiggling more slowly. The bigger the temperature difference, the more energy the electrons have to flow. However, the voltages TE materials produce are absolutely tiny – not much use in a typical car. To make them practical, scientists developed thermoelectric materials that can conduct positively charged particles (remember, electrons are negatively charged). By connecting lots of these pairs of materials together, we can produce voltages large enough to do something useful with. In the exhaust pipe of a typical family car, we can easily see temperatures upwards of 300°C (572°F). By placing an array of TE devices around the hottest part of the exhaust and cooling the other side with air or water, heat can be harvested and turned into electricity.[*]

Now, there is one killjoy in this situation, and I'll hand over to, arguably, the greatest philosopher of our time to describe it. Homer J. Simpson once said, 'In this house we obey the laws of thermodynamics!' And it's these laws that limit the performance of these devices. TEs depend on a temperature difference, but if you think about how quickly a cold cup heats up when filled with a hot drink, you can probably identify the issue. Hot surface + cold surface = warm surface (thermodynamics in action!). Once that temperature difference disappears, so too does the electricity production. As well as that, the materials themselves – including the rather exotic-sounding bismuth telluride – are expensive and difficult to make. But even with these limitations, cars and trucks equipped with thermoelectric devices would deliver significant fuel savings, so car companies are hugely interested. In fact, in 2012, BMW demonstrated this proof of concept, by installing TE devices on the exhaust line of an X6. With one side of the materials nice and hot, the other was cooled using water from the car's radiator. Their prototype produced about 500w of electricity from running on a motorway. It was used to power various systems throughout the car (lights, audio, warning systems) and so reduced fuel consumption by 5 per cent.

---

[*] Spacecraft and planetary rovers have used TE materials for decades – Mars Curiosity is almost entirely powered by a radioisotope thermo-electric generator. Rather than an exhaust pipe, the heat source is a tiny block of a radioactive material.

If we can find a way to slow down the flow of heat through a TE material, we'd give engineers a little longer to harvest electricity, making everything more efficient.[*] This would have a huge impact on the automotive industry, and anywhere else that heat is lost. Possible solutions include moving to nanostructured materials called skutterudites, which trap heat inside their cage-like crystal structures – it's these materials that are the current record-holders for efficiency.[†] But finding a way to make them on an industrial scale is still some way away. Fingers crossed, though.

It's not just heat that cars could harvest in future – we could even steal energy from friction. Now, friction is very important for road vehicles. If you've ever had to drive on an icy road, you soon miss that 'sticky' force we all take for granted. But while wheel-based systems like cars and trains rely on friction (and we'll talk about this more in Chapter 6), it is also a major source of energy loss. According to the US Department of Energy, 5–7 per cent of the energy in a vehicle's fuel is lost to so-called rolling resistance between the tyres and the road surface. And where there's a challenge, there are researchers trying to solve it.

One particular group of Chinese and US researchers used the triboelectric effect to harvest electricity. Have you ever rubbed a balloon in your hair and made it stick to the wall? Then you're already experts, but let's talk about it anyway. A triboelectric generator is nothing more than two carefully chosen materials that are repeatedly brought together and separated. Every time they make contact, electrons move between the layers, and this voltage can be collected and used to power other things. In what must be my favourite set of pictures in a serious scientific paper, this same group of researchers rigged their triboelectric generators to the tyres of a toy jeep (think Action Man). When the jeep was rolled

---

[*] Right now, bismuth telluride transforms just 4 per cent of the harvested heat into electricity, when its 'hot' side is at 250°C (482°F).
[†] Skutterudites got their name from the mine in which they were first discovered – Skutterud in southern Norway

across tarmac – a material that likes to give electrons away – the device on the wheels collected them, and used them to power the jeep's LED headlights. While still at the very earliest of stages, this work also suggested that the effect would be even greater on a heavier, life-size vehicle (because heavier car = higher friction between the tyre and road surface). I am totally cheering for them on this! Sing it with me, 'Rolling, rolling, rolling...'

So now that we can capture waste energy in many forms, what are we going to do with this extra electricity? We'll use it to power the on-board sensors of course! And to figure out how, let's talk about one more option – piezoelectric materials. We last met them in Chapter 4, embedded into paving stones. In cars, they will be used in self-powered sensors. Wireless tyre-pressure sensors are already installed in the air valves of many vehicles, but they need batteries to power their data transmission. Tomorrow, thin strips of piezoelectric materials could do the same job. Vibrations in the tyres (from the motion of the car) could bend the piezo, shifting electrons from their usual position and providing an electric charge that can be stored and tapped into to power the sensor. Current research is investigating even smaller devices made with cheaper and more robust materials. It's all looking rather exciting.

Materials that harvest the energy we throw away may only produce tiny amounts of electricity, but it is more than enough for today's automotive sensors – they've never been so efficient. A wireless transmitter needs no more than a few hundred microwatts (0.0002W), and all of the materials we've discussed here can provide that. So now that we have the means to power a new generation of sensors, it's time we talked about what we need to sense.

Sense

Hopefully by now, we're all in agreement on the need for lighter, cleaner cars. And we have some idea of how we can power the app-enabled future. High-end sensors and in-car tech used to scream luxury, but they've become so cheap that they're now everywhere. Most vehicles already have between 80 and 100 dedicated sensors on board, monitoring everything from engine temperature to fuel chemistry, but within the next decade, that figure will double (at least). And given that in 2014, the global market for automotive sensors was estimated to be worth somewhere around $20.27 billion (£14.49 billion), it's a pretty good sector to be in.*

The majority of these new sensors will focus on safety. Video cameras and electric steering that detects and adjusts when a driver is straying out of lane are already standard in lots of cars. They're frankly boring compared to what tomorrow's sensors will be able to do. They'll be smarter, and will form part of the fabled Internet of Things (still to come in Chapter 7). Allowing cars to communicate with each other means that you'll be informed about traffic accidents as they happen, or whether parking spaces are available within a defined area. Car-to-grid communications could be used to measure the power demands of electric vehicles, to monitor traffic or to help city officials identify problem traffic areas. IBM carried out a small pilot on this very topic back in 2013. The location was the Dutch city of Eindhoven, a major European transport hub. In the trial, 200 cars were equipped with a range of sensors, measuring their location, as well as braking and acceleration data. Their trial showed that even this low level of data from vehicles helped traffic engineers to be more responsive to changes in traffic, 'reducing congestion and improving traffic flow'.

But there are also plenty of less critical, more fun ways to use sensors and other technologies in cars. In reality, there is no end to the optional extras that will become standard in the cars of the future. To be honest, the real challenge for manufacturers is to make sure the tech doesn't become a dangerous distraction to the driver. Keeping it completely out of the car just isn't going to happen. Anyone attached to their phone is familiar with the separation anxiety of being

---

* Figures like these should always be taken with a pinch of salt, but having looked around at various market reports, this seems plausible.

'disconnected', even for a short while.* A study published by management consultancy McKinsey in 2014 suggested that one in eight people would immediately rule out buying a new car that didn't offer internet access. Yes, really.

The key for tomorrow's app-enhanced cars is to reduce the cognitive load on the driver; allowing them to keep their eyes on the road and hands on the wheel, while still offering a way to stay online. Products like Google Glass (essentially a wearable smartphone) may be one option, but I want to focus on in-car tech. Some high-end vehicles today can project basic information such as speed and mileage onto the windscreen, as a primitive heads-up display. In the future, we'll go much further than this, all the way to augmented reality dashboards, allowing us all to channel our inner *Top Gun* fighter jocks (I'm Maverick, you can be Goose). These displays will identify objects and tell the driver how far away they are. They'll also allow you to overlay your GPS map onto the road you're driving on. They'll read signs for you, analyse the road surface, and monitor your vital signs to check for drowsiness. Mechanics will be able to interact remotely with the display to identify problems with the car.

The technology needed to do all of these things (and more) already exists, and they're slowly beginning to come together to make that vision a reality. The thing that strikes me about all this is how far even today's cars have moved from their original form. They're less like a high-speed cart, and more like a data centre on wheels. We, their human operators, still have control, at the moment. Lots of these sensors 'learn' about our driving behaviour, with some already taking limited responsibilities, such as active parking. So are we moving towards a day when the driver is no longer needed?

Driverless
Driverless cars no longer seem like quite the crazy, sci-fi idea they once were, but before we start discussing them, I must

---

* I jest, but there is a lot of ongoing research into the implications of tech addiction.

begin with the following proviso: there is almost no point in me trying to predict the *potential impact* of driverless cars. I mean that very seriously. Right at the start of our journey together, we talked about the elevator. Before them, living several storeys up was a pain. After them, penthouses appeared at the top of every building. More practically too, without the development of the elevator, we could never have dreamed of 'scraping the sky' with our homes and offices. Elevators didn't just change the way we did things – they built the city, and no-one could have predicted that outcome. That's kind of how I feel about driverless cars – they could be *so* transformational, that we can't even begin to imagine the future they could bring. But we can at least talk about the tech behind them.

The allure of automation is huge for the car manufacturer and the consumer alike, and don't we all dream of being able to kick back, relax and enjoy the ride? Despite the hype, I'm pretty confident that this particular level of autonomy is still some years away. It's definitely going to happen, but realistically in the short term (10–15 years), the tech is mainly going to augment human drivers, rather than replace them (us) outright. This message was reiterated at the 2015 Mobile World Congress. Carlos Ghosn, chief executive of Nissan-Renault, said it would require a decade for regulatory issues related to robotic cars to be put in place, but that within five years we'd see cars that could manage motorway driving without human intervention.

BMW, Mercedes-Benz, Nissan and General Motors have all launched cars capable of driving autonomously in limited situations, but it was Google's announcement in 2013 that sent media interest into a frenzy. Working with Toyota, Lexus and Audi, they are working to develop the world's first fully autonomous car. Software along with a host of sensors and hardware has been retrofitted to various vehicles, and as of May 2015, Google's cars had logged over 2.7 million kilometres (1.7 million miles).* This is impressive, but if we look beyond

---

* This number combines manual and autonomous driving, for their fleet of 20+ vehicles.

the headline figure, we see that it is dominated by their performance on private roads or motorways, rather than in a built-up urban environment. And this is partly because on the busy, ever-changing streets of a city, there's a lot going on. Drivers need to remain alert to their surroundings, absorbing lots of data and making many more decisions than when driving on the motorway. Computers are remarkable, but they aren't truly 'intelligent'. Even the best supercomputers can only make decisions based on the rules they've been taught by a human programmer – for example, 'if you see that this light is red, stop.' So when they come across something entirely new, say a broken traffic light or the siren of an emergency vehicle, they don't always have the answers. For black and white decisions, autonomous cars are kicking our asses, but for the 'grey' moments in city living, nothing beats the human brain. For now.

There is a real momentum behind driverless vehicles. As of January 2016, two low-speed, self-driving buses were shuttling passengers along a 6km (3.8-mile) stretch of public road in the Dutch city of Wageningen, and other urban pilot schemes are well on their way. But how do driverless cars even see where they're going? The main option is LIDAR – which either stands for **LI**ght **D**etection **A**nd **R**anging or is a combination of *light* and *radar*, depending on who you ask. The key thing to note is that it uses lasers to 'range' objects. LIDAR can do this thanks to the relationship between distance, speed and time – if we know any two of those things, we can measure the third. We know how fast laser light travels, so if we measure the time it takes for the beam to leave the transmitter, bounce off an object and return, we can calculate the distance to that object.* Previous automotive LIDAR systems were mounted onto car

---

* The speed of light in a vacuum, referred to as c, is a constant and is 299,792,458 metres per second (or about 186,300 miles per second). It travels slightly slower in air, but its value is known – it is about 299,705,000 metres per second (or 186,200 miles per second).

bumpers, and they were used solely for cruise control. But the new breed of LIDAR is an entirely different beast; Google's car uses 64 laser beams which can take 1.3 million measurements per second. And because it is mounted onto a car roof, it can build up a 360°, 3D map of the area.

Could the average city-dweller afford to buy a laser-scanning car? At the moment, Google's LIDAR is extremely expensive – upwards of $50,000 (£35,700) in 2015. But commentators expect the next-generation version to be considerably cheaper. And frankly, it'll need to be. Driverless cars will need other tech too – high accuracy GPS, short-range radar sensors and video cameras… it all adds up. And that's before we even talk about the beasty on-board computer that driverless cars will depend on. Millions and millions of data points will be generated every second that the car is on the road, and they need to be processed almost instantly, to make decisions in real time. No standard computer could manage that – honestly, this could be the largest expense.

However, the biggest challenges facing driverless cars may not be tech-related. The Transportation Research Board of the US hosted a week-long symposium in 2015 that specifically looked at the future of transport. Experts from across the world presented their research and debated the ins and outs of various options. Throughout the event, one common theme emerged: the *logistics* of driverless cars. There are behavioural concerns – for example, how comfortable would you be sitting in the driver's seat of a car without being able to control it? When there's a mix of old and new vehicles on the road, will that change the way humans drive? And what about the legal and moral ramifications – who takes the blame if a driverless car is involved in an accident? How will they make decisions in worst-case scenarios, such as driving into a wall versus hitting a pedestrian? The infrastructure needed to 'house' driverless cars is far from up to scratch either – can a computer process a street sign that's covered in graffiti? The questions are endless, and go way beyond the science, engineering and technology needed to actually build autonomous cars.

One big thing that's often left out of the debate is security;
I don't mean to be a complete scaremonger, but wherever
there's a computer, there's a way to hack it. I think Roger
McKinlay, President of the Royal Institute of Navigation,
spoke for many when he told me, 'I for one would not feel
comfortable entrusting my life, or those of my loved ones, to
a driverless car until the security was absolutely watertight.'
And it's not an unreasonable concern. In 2015, Andy
Greenberg, Senior Writer for *Wired,* worked with two digital
security researchers (hackers) to carry out an amazing, if
scary, experiment. While Andy was driving along the freeway
near St Louis, the hackers accessed the car's on-board
computer. Once they'd got into the system, they could switch
on and off the air-con, radio and wipers, and eventually
managed to stall the engine, leaving Greenberg stranded
(briefly). But the hackers weren't in the back seat of the car;
they were more than 10 miles away and sent all of the
commands via the internet. As our cars continue to become
more connected, security concerns like this will only grow,
but panicking rarely leads to good science, so let's not do that
here. The hacker duo involved, Charlie Miller and Chris
Valasek, have spent many, many years working with car
companies to poke holes in their software. It's only been
through thousands of hours of focused coding that they
managed to access so much data for this one manufacturer. So
for now, the good guys are miles ahead of any potential
'baddies'. But we shouldn't get complacent – security will be
an issue for our connected future, and the tech will need to
stay one step ahead.

The ability to map the world around you and identify
hazards is definitely one huge step on the road to driverless
cars, as is our growing understanding of the possible security
issues. But there's much more to driving than that. We interact
with our vehicles in many ways: we check our wing mirrors,
change our speed using pedals and sticks, indicate when
turning and steer using a wheel in front of us. If we take the
driverless car to its natural conclusion, we won't need these
inputs. Right now, all prototype driverless cars include this

stuff, because human drivers always take over at some point. This will remain the case for some time yet – until every single car on the road has the ability to drive autonomously, no-one's going to take away your steering wheel.

While our relationship with cars is changing at an unprecedented pace, it is clear that we're still far from building the flying cars that 1980s kids' TV promised. However, we're at a real tipping point in the history of the road vehicle, and it's been brought about by excellent science and engineering from researchers all over the world. Never has the future of cars seemed more exciting, or less easy to predict.

CHAPTER SIX

Loco

Deep beneath the urban soil, a vast maze of concrete tubes can be found. Every few minutes, metallic boxes full of intelligent lifeforms speed through them, stopping regularly to change their cargo. In cities across the globe, this system runs 24 hours a day, seven days a week, and is often considered the lifeblood of the urban sprawl. In London, it's the Tube; in New York, the Subway; and in Paris and Tokyo, the Metro. Whatever nickname you give to underground rail systems, they've had an indisputably important role in shaping the world's great cities.

So, in this chapter we will immerse ourselves fully in the world of trains, and in the infrastructure that enables them. Many of us take to the tracks on a daily basis, but more than anything we've met so far in *SATC*, underground train systems are out of sight, out of mind. We'll uncover some of the most impressive engineering you've never thought of, and if you've ever wondered how trains navigate, or why leaves on the track cause havoc, you'll find the answers in here too.

Before we launch into it, a warning. Prior to researching this chapter, I was unaware that I was a closet train and tunnel nerd. Writing about them has unleashed a monster of epic proportions, and it's possible that reading about them will do the same. With that in mind, let's agree on some of the vocabulary.

Today

Much as I've grown to love anything rail-related, I could never be accused of being a purist. For the purposes of this chapter, when I say trains, I pretty much mean anything that travels on tracks. This could include trams, light-rail trains, underground trains, heavy-rail passenger trains and freight trains. While not all trains are created equally, many share a similar setup, so that's my excuse for grouping them together.

Locomotives are first on our list, and they tend to be used by larger trains. The word originates from the slamming together of two Latin words: *loco* ('from/to a place') and *motivus* ('moving'). Stuck onto the front of the train, the locomotive is its engine. It provides the energy needed to get it moving and to pull the cargo and passengers along behind it. As we'll discover, this isn't the only option for powering a train along its tracks – those that carry us under the city streets run on an entirely different engine. But without the invention of the locomotive, trains may never have made it underground, so let's start there.

Steam

We can't talk about trains without mentioning steam (or 'gassy water' as I once heard it described). It has been used to move stuff for centuries, but it was during the Industrial Revolution that it was first applied to big, practical locomotive systems. What makes it special is pressure. The molecules in liquid water are closer together than those in steam – as you increase the temperature, the bonds that connect them to each other stretch, and the molecules spread out. This means that normally, steam takes up more space than water.* But instead of letting the molecules spread out, we seal it into a strong container so that it can't escape. At 100°C (212°F), the H_2O molecules inside the container have huge amounts of kinetic energy, and they bash wildly into each other and against the walls. If this build-up of pressure can be released carefully, the energy of the gas molecules can be tapped into, and this is what powered steam locomotives of old. Coal was burned to produce heat energy. That heat boiled the water in the tank, producing steam, and this steam was piped into a

---

* When boiled, a 1 cubic centimeter droplet of liquid water would produce 1.7 litres of steam – a difference in volume of 1,700 times.

piston and cylinder setup similar to that used in a car engine. The in-and-out motion of the piston turns the crankshaft, driving the rotation of the wheels, and moving the locomotive along the tracks. The recognisable sound of a steam train – that intermittent *chuff-chuff-chuff* (that evokes misty-eyed memories in many) – is really the result of the pistons moving in and out of their cylinders; it is the locomotive breathing.

The first full-scale steam locomotive was built by British engineer Richard Trevithick in 1804, but George Stephenson was really the one who made them practical. His first locomotive could haul up to 30 tons of cargo uphill at 4mph. While that might sound slow, it was more than could be towed by 10 giant shire horses, making it a revelation at the time. There was a downside, though. Steam engines work by external combustion – in other words, energy is released in one place (in the boiler) but is required elsewhere (in the wheels). Anywhere you have to move energy, you also lose energy, and in fact, only 10 per cent of the chemical energy in coal helps to turn the train wheels; the other 90 per cent is used to heat the water. On top of that, producing steam takes time, a fact familiar to those of us who stare at boiling kettles when under-caffeinated. Much as I love steam trains, a more efficient, lower-maintenance engine was sorely needed.

The 1940s brought with them a surprisingly forward-looking solution to the problem of train power: a hybrid engine that combined diesel and electricity. Diesel-electric locomotives use a constantly running diesel engine to spin an electrical generator at incredibly high speeds. In turn, the generator powers the motors connected to the wheels, allowing the train to move. Hybrid locomotives are robust, reliable and considerably more efficient than steam engines. Despite being around for 70 or more years, they remain in widespread use in suburban and intercity train networks. One place you won't find them is Tube tunnels, and this is due to their emissions. Ultimately, these locomotives get their power from a big tank of diesel on board – it may not be directly used to turn the wheels, but it is being burned, and so produces nitrogen oxides and carbon dioxide. In high

concentrations, these gases are damaging to human health, so
using them in an enclosed space would be really stupid. Our
underground trains needed something else.

Sparks
Showing themselves to be far more futuristic than cars, the first
fully electric locomotives were developed 180 years ago. By the
second half of the nineteenth century, cities were booming,
and there was a growing need to fit transport options into an
already busy landscape, so underground railways were a must.
London fully embraced these two trends as early as 1890,
building the world's first deep-level underground 'tube'
railway, which was populated by electric locomotives hauling
simple carriages. Just a few years later, those trains were replaced
by the creatively named electric multiple units (EMUs) – self-
propelled carriages that can run on electricity, no locomotive
required. Today, EMUs are widely used in cities across the
globe – Paris, Moscow, New York, Rio de Janeiro and
Copenhagen, to name just five. And the reasons are clear.
Electric rail vehicles are reliable and quiet, and they don't
produce any emissions (locally at least), which is particularly
important in tunnels and built-up urban areas. They are also
remarkably efficient; an electric locomotive could produce
almost three times as much power as a diesel locomotive of
similar weight. But we haven't yet answered a key question:
where do trains get their electricity from?

Just like our homes, the rail network sources its electricity
from the grid, transmitted via high-voltage lines. Once it gets
to the network, there are three main options for distributing
the electricity to the trains themselves:

1.   on-board energy storage systems, such as batteries;
2.   an overhead wire that the train connects to; or
3.   an extra 'live' rail that has direct current flowing
     through it at all times.

Batteries aren't visible to passengers, but you've probably
noticed at least one of the other two options on your rail

journeys. Overhead wires are best suited to tram and intercity services, whereas the more compact 'third rail' option is preferred for underground trains. The role of the third (or conductor) rail is to ensure that the electricity is always directly available, so it's installed alongside, or in between, the pair of running rails. Keep an eye out for it when you're next on an underground train. If you're in Milan, London or certain areas of Paris, you might be extra lucky – those systems have four rails. You might remember that electricity always needs to travel in a loop. You can't just send out a bolt of it; it needs to come back too. On most trains, the current goes out on the third conductor rail, and returns on one of the two running rails. But in the early days of a few underground systems, the return current seemed to get lost. Engineers realised that sometimes, it didn't take the route back along one of the main rails. Instead, the current flowed through the iron that was used to clad Victorian train tunnels, or along the iron water pipes that ran beside the rails. Given that neither of these was designed to carry electricity, they were badly damaged. Adding a dedicated fourth rail solved the problem – being made of a steel alloy that is more conductive than cast-iron pipes, it happily carried the return current to whence it came.

Conductor rails carry lower voltages than overhead wires, although 'low' is a relative concept. London Underground's system provides 600V of direct current to their conductor rails, still enough to be potentially lethal. In addition, the use of conductor rails sets a speed limit on trains of about 160kph (100mph) – above this, the metal contact blocks (called pickup shoes) can lose contact with the rail and result in a drop in power. The pickup shoes do occasionally lose contact, albeit very briefly, at track junctions. This is the cause of the very bright, blue-white spark you sometimes see near the third rail, as well as flickering carriage lights. For high-speed trains, overhead wires are the better choice.

Track
While it's clear that you can't have trains without tracks, the concept of using tracks to transport goods predates the

locomotive by some way – about 2,400 years in fact. Grooves cut into limestone slabs were used as tracks in Greece in 600 BC. Wheeled vehicles pulled by animals along these tracks transported boats across strips of land. By the 1500s, the picture had evolved only slightly, with miners in Germany using wheeled carts to transport ore between parallel planks of wood. The Industrial Revolution introduced the need to move goods over long distances, as cheaply as possible. Many experimented with cast-iron and wrought-iron rails, but for the next 150 years, they remained limited in their use. Just as in our buildings, it was steel that made them practical. The plummeting cost, ease of production and mechanical properties of steel made it a realistic solution, and even today, it's used in the overwhelming majority of rails.

But why do tracks make more sense than roads for shipping dense goods like coal? It's partly due to rolling friction. We know that friction can be helpful sometimes – soft, patterned car tyres can easily grip on to a rough tarmac road. In contrast, train wheels are very stiff and relatively smooth, and they roll along a narrow, rigid steel rail. It's not that there's no friction between train wheels and rail – far from it – but it is lower than that between tyres and the road. So for moving stuff around, rail is the more energy-efficient (and therefore cheaper) option. As in many engineering challenges, friction is a compromise: too much and the train grinds to a halt, too little and it can't move at all. And friction (technically, the loss of it) has a dark side. Another one of the questions I was asked while writing *SATC* was **why do leaves on the line cause rail systems to grind to a halt?** That's not merely the perception of a frustrated commuter; in the UK in 2013, leaves were directly responsible for 4.5 million train–passenger hours of delays, and it's mostly because they alter the friction between the rail and wheel. When wet leaves are compacted under the weight of a train, they form a hard, slippery, wax-like coating, which reduces friction so much that the wheels lose their grip. This means that trains either can't get going, or if they're already moving, they struggle to stop. Today's solution is to fit trains with nozzles that either blast jets of water onto the tracks to clean them, or coat them with sand to claw back some much-needed friction. Tomorrow's solution is much, much cooler, but I'm afraid you'll have to wait a little while longer to hear what it is.

Where you are in the world tells you something about your train track beds. Most of those in use in Europe and the Americas consist of steel rails supported on either timber or toughened concrete sleepers. The rails have a unique profile, with the most common called 'flat-bottomed' – this shape allows them to sit level. The rail–sleeper assembly itself lies flat on a thick layer of crushed stone (called ballast), which helps to support the load of a passing train, as well as draining water away from the steel rails. The tracks of most high-speed trains across Asia don't use ballast at all; instead the rails and sleepers are directly attached to a large concrete slab. Either way, it's all about spreading the load evenly, while keeping the track as stable as possible to optimise the train's running.

Trains cannot run on rails alone. They also need wheels, and they are a thing of beauty, honestly. Almost all of the earliest track systems were based on a wheel that was set inside a guide, be it a groove in the road, or wooden planks. But any train today very much 'rides the rails', and that's thanks to clever design.[*]

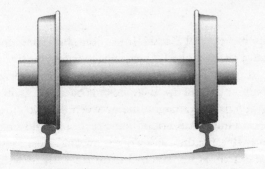

Figure 6.1 As shown in the somewhat exaggerated image above, train wheels are designed to hug the rail.

---

[*] Tram rails look somewhat different to those used on metro or intercity trains – their wheels sit inside a curved steel rail in order to ensure that the tracks are flush with the road surface.

If you were to look at a pair of steel train wheels in profile, you'd see that they aren't cylinders – rather, they're conical, slightly wider at the inside edge than the outside, which helps with guiding. They also have flanges attached to their inside edge – these are discs with a slightly larger diameter than the wheel itself. Flanges are an important safety feature, the very last resort to stop a train from derailing. You might have spotted too that the rails are angled slightly inward. This combination of rail and wheel profile means that the rail 'head' and the wheel can be in excellent contact, even when the track curves around corners. Some metro systems, such as the one in Mexico City, use rubber-tyred wheels rather than steel ones – these rattle around a bit less, so tend to be used on tracks built on unstable soils. But no matter what material is used, train wheels are not magic, and drivers must still limit their speed. The maximum speed that a train can safely travel around a bend will depend on how sharp the bend is, the weight of the train, the shape of the rail and the steepness of the route. Hills are rather tricky for trains as well, and to explain why, let me tell you a story.

Roll

It's 1841 in Imperial Russia. In a room, bickering engineers surround Tsar Nicholas I. They've been trying his patience for weeks, threatening his vision of a rail link between St Petersburg and Moscow. Exasperated, Nicholas grabs a ruler and quickly draws a straight line between the two cities of his empire on a huge map, not noticing that one of his fingers was in the way. Because the Tsar's word is law, engineers constructed the line exactly as he sketched it on the map, including one curious deviation. And indeed, until 2001, a semi-circular kink in the line could be found near the city of Verebye, all because of the Tsar's finger.

Oh how I wish this legend was true. I'm afraid the real reason for the kink is far more prosaic – it was down to the pulling power of trains. At the time, even the best locomotives couldn't cope with hills, and Verebye had the steepest slopes of anywhere along the 650km (404-mile) route. If the engineers had built the tracks on the hill, trains coming from Moscow would have needed four locomotives to make the climb. And those en route from St Petersburg would gather so much speed on the way down the slope that they wouldn't be able to stop at the next station. The only solution was to construct a detour route that gradually worked its way up nearby shallow slopes.

But why are trains at the mercy of steep slopes? Let's have a think. We've all had to carry shopping bags or a heavy rucksack around, and even if you're weighed down while walking on a flat surface, instinctively you know that once you start uphill, the going will get tougher. Physics is to blame for that. You may not be aware of it, but every time you lift an object or climb a hill, you are winning a quietly fought battle. Moving things along a flat surface really only requires you to overcome friction. But moving things up a slope pits you against yet another foe – gravity! The thing about gravity is that it is reliable – it only acts in one direction, down towards the centre of the Earth. For things moving across the surface of the Earth, say a train on a track, gravity doesn't really influence its motion – the locomotive just happily pushes or pulls it along. But once you deflect from a horizontal surface, gravity begins to take notice. Going up a hill involves going against gravity. And the steeper the hill, the harder you (or a train) must work. Slopes can be defined in lots of ways, but because I like angles, we'll go with that. In the sketch below, you'll see that the slope of this hill is labelled α. When $\alpha = 0°$, you're on a level surface, but anything above that value means that you're on a slope ($\alpha = 90°$ for a sheer cliff face).

If you were to guess the maximum steepness that today's fancy trains could cope with, what would you say? 30°? Maybe 40°? I'm afraid you'd be WAY off. The steepest slope that friction-based trains can manage is just 4°, but even that would be only over a very short distance. This is because of the other force acting on the train: friction, which we've

Figure 6.2 When trains move along the horizontal, gravity (the big arrow) doesn't really have an effect on the train because it is balanced out by the 'reaction' forces from the rails (skinny arrow). On the slope, the train has to do more work against the force of gravity – the forces no longer balance.

already talked about. As slopes get steeper, the lovely adhesive forces that typically help a train to move along the tracks are no longer enough to keep wheels in contact with the rail – the train literally loses its grip and can slip back down a slope. Add that to the fact that even a tiny slope makes a huge difference to the train's 'pulling power': a 300-tonne train will have to work almost twice as hard to climb a 1° slope than to travel along a flat surface.

Of course, the opposite is true for a downward slope – the steeper the drop, the more rapidly the train moves (because gravity is helping it). While this sounds useful, in reality, a steep drop can be even more troublesome for a train; there's a reason that the phrase 'runaway train' has stuck around. So while locomotives have become much more powerful since the era of the Tsar's finger, gravity will always limit what friction-based trains will be able to do.* However, as with all engineering difficulties, it is possible to transform the slope affair into a solution.

To learn more, I spoke to some of the world's finest tunnellers: the Crossrail team. Crossrail is Europe's largest construction project, which is adding 42km (26 miles) of

---

* Thanks to more powerful trains, engineers did eventually remove the Verebye kink.

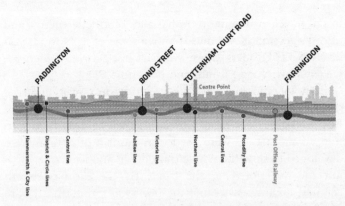

Figure 6.3 A cross-section of part of the Crossrail route reveals the clever use of slopes to slow down and speed up trains. (Credit: Crossrail)

huge new tunnels under London.* A section of the tunnel schematic shows that it rises and dips through the city soil. But there is nothing random about this profile. The route isn't just defined by the location of existing stations or the rock type (although both are very important). There are other interesting features common to all of the stations.

It might not be immediately obvious in this (not-at-all-to-scale) image, but the track has been planned so that it's always lower between stations than it is in the stations themselves. Think about the behaviour of a train travelling east into Tottenham Court Road: as it enters the station, it has to climb a small slope to reach the platform. On leaving the platform, the train drops down another small slope. Crossrail uses gravity to either slow trains down (by forcing them uphill on the way in) or speed them up (by forcing departing trains downhill). Other urban train systems use the same approach, because it not only makes for a more pleasant passenger experience, it also saves a lot of energy in the brakes.

Stop

If you're a frequent user of any underground train system, you'll know just how much stopping and starting is involved

---

*You'll hear a lot more about these tunnels soon. I was lucky enough to wander through sections of them with the engineers themselves. Yes, I'd be jealous if I were you too.

in rail travel. A moving train has a lot of kinetic energy, and in order to stop it, we must remove some of it. Now, as you lovely *SATC* readers know, we can't just get rid of energy – we can only change it into another form.

Early locomotives were relatively low in power, and so a simple handbrake system could be used to slow them down. But as engines got more powerful, they became harder to stop, so alternatives were needed. A number of systems were developed to slow the whole train down at once, rather than just the locomotive at the front. By the early twentieth century, air brakes became the norm on trains, and they're still in use across the world today. Also known as pneumatic brakes, they use compressed air to rapidly push pads or discs onto the train's wheels, forcing them to slow down by increasing friction. Air brakes are also the source of that hissing sound that you hear from braking trains – puffs of compressed air are released to equalise the pressure in the brakes. And just as an aside, the heat produced by these brakes also warms up the surrounding air. So much so, that it has to be considered when designing tunnel ventilation.

While friction-braking works very well, it has its limitations. If you used these brakes to slow a train from its top speed down to stationary, you'd quickly wear out the brake-pads. So instead, they're reserved for only the final bit of braking; as a train enters a station, for example. Before that, it is regenerative braking that slows trains down, similar to that used in Chapter 5's electric cars. These systems are found on underground trains in Los Angeles, Auckland and Buenos Aires, among others. By using the 'brake' to change the connections on a train's motors, they stop turning the wheels, slowing the train, and instead produce electricity that can be used elsewhere. In all cases, this regenerated electricity can be returned to the power line via the conductor rail. How it then gets used depends on the city we're in. If the train line is a busy one, with small gaps between scheduled trains, the electricity produced by one helps to power the very next one that comes along. If it's a less busy line, some of the electricity can be stored in batteries on the train. If they use direct current, electric brakes can generally slow the train to about 8kph (5mph). If they use alternating current, they can slow the train to nearly a full stop.

So we are now experts on how trains move, climb and stop. But what about knowing where they are and how they avoid congestion? The first bit is pretty easy – some train systems use electrical circuits built into the rails. Effectively, if a train is present on that section, a switch turns off, and if there's no train, the switch remains on. Another option is a pair of axle counters – as the name suggests, each time a train's axle (or pair of wheels) passes by, the device counts it. By installing them in pairs along a section of rail, and doing a bit of maths, you can measure the direction, length and speed of a train. Knowing where trains are is only part of the challenge, though. In built-up urban areas where station platforms are almost always busy and there are many trains on the lines, control is really the key thing, and that leads us to signals.

We've all been on a train brought to a halt by 'signalling problems', but how many of us know why they occur? Traditional signalling systems are based on the very simple idea that trains won't collide if they're kept at a safe distance from each other. To manage this, railway lines are separated into discrete sections called 'blocks', which have signals (typically, coloured lights) at each end. When the signal changes, it indicates to the train driver that they can pass to the next block. These signal-controlled blocks of track can be any length, from a few hundred metres to tens of kilometres long, depending on how many trains use the route. Now, like most things on urban rail systems, signals depend on electricity. So if there is a problem with the power supply (and that really is almost always the issue), signals simply won't work, and train drivers are not allowed to pass them.[*] The short-term fix for this is to install 'surge-protected' systems, to ensure that the power supply can't be interrupted. The

---

[*] As an aside, this would also be true in the case of fully automated trains, so maybe don't be so tough on your driver.

longer-term approach might even solve the other issue with 'fixed block' systems – speed.

Today's trains are much faster (and the rail network considerably busier) than when blocks were first designed. Higher speed means longer braking distance, but this isn't really reflected in the length of the blocks – basically they're too short for our speedy trains. This means that trains on busy urban lines spend a lot of time braking. Add even a small delay into this, and we see a cumulative effect on the whole system (remember the jamitons on the motorway?). Because the blocks are a physical part of the rail infrastructure, they also limit how many trains can use the track at one time. We'll come back to the solution to this shortly, but first, we need to explore beneath our feet.

Dig

We can build trains and we can lay tracks, but in most cities, trains are hidden out of sight. If we're going to hide trains, we have got to dig, so geology and tunnels are up next. Please note that while I promise to try to avoid going full tunnel-nerd on you, I may not manage it…

There are many ways to dig a tunnel, but all of them require an in-depth knowledge of the rock type. Depending on your city, the ground can be made up of anything from tightly packed sand to volcanic rock, so long before a tunnel's route is finalised, geologists are drafted in to map the rock in detail. A geological survey is a complex task that includes lots of different techniques. For a large-scale civil engineering project, it would be typical to start with ground-penetrating radar, to check for the presence of any buried utilities. According to geologist Dr Alan Baxter, 'You would also look at the geotechnical aspects, identifying joints and faults to identify planes of weakness.' The next step would be to collect samples from boreholes to find out the detailed composition of the rock, along with its water content. Even after you have fully mapped the rock type in the proposed digging area, tunnelling projects are still full of surprises – during the first construction phase of the Mexico City Metro, workers unearthed an Aztec pyramid. In London, Crossrail workers found everything from ancient flint tools to Roman hairpins. In digging tunnels for the LA Metro, fossils dating back 16.5 million years were discovered, and for Istanbul, the biggest underground treasure was a Byzantine shipwreck. All of this really goes to show that when you tunnel under a city, you also tunnel through time, catching a glimpse of what once stood on that very spot. So these days, urban tunnelling teams always include skilled archaeologists, who work closely with those doing the actual digging.

Today's tunnels are broadly dug using two main approaches: cut-and-cover or deep bore. Each has its advantages and disadvantages, but fundamentally, the decision on which to use comes down to three considerations – geology, money and the function of the tunnel.

On the subject of cut-and-cover tunnelling, used extensively in the NYC subway system, I was lucky enough to interview Ailie MacAdam, global rail boss of Bechtel. She has worked on some of the world's most famous construction projects, including Boston's Big Dig. 'Really, for this sort of tunnel, what you are creating is an underground box, with parallel sides set at ninety-degree angles.' In Boston, they used a 'top-down' approach. First, thick concrete panels (called diaphragm walls) were inserted into the ground along the proposed tunnel's route – these would eventually become the walls of the box-tunnel. Then all of the material between these walls was removed, forming a trench. As you'll no doubt remember from building forts and hideaways in your living room, walls don't stand up by themselves, so in cut-and-cover tunnels, they are supported by enormous temporary struts. Slabs can then be added to the bottom of the trench, followed by a roof, and once that's done, life can return to normal above ground while work carries on under it.* This wasn't just one simple tunnel, though. The Big Dig involved

---

* And as I can attest to, the default mood of anyone attempting to cross a city full of construction works is Grumpy with a capital 'G'. Don't underestimate how important avoiding large-scale traffic and pedestrian disruption is to city planners!

creating an eight-lane underground expressway through Boston. Because of the tunnelling process used, they managed to do it without shutting down an existing elevated highway that ran along the same route. The *Washington Post* artfully described this process as 'like performing open-heart surgery on a patient running a marathon'.* Engineering never fails to blow my mind.

Before we talk about tunnel option no. 2, let's just remind ourselves of something. Like the bridges we met in Chapter 4, tunnels are all about balancing forces. Only this time, as well as supporting itself, the structure needs to support all of the soil, rock or water that surrounds it, as well as any existing infrastructure above it. Tunnels have to withstand tension (pulling forces), compression (squeezing forces), shearing (sliding forces) and torsion (twisting forces). Which is why heavy, strong materials such as steel and concrete are used.

Coping with these forces comes easier to some shapes than others. A cut-and-cover tunnel is a box, with flat sides and pointy edges. This means that the weight of the soil surrounding it is not distributed evenly. Think about a person standing on your bare foot while wearing slippers, and then again wearing a pair of stiletto heels. Which one hurts more? The person's weight remains the same both times – what changes is the area over which their weight is spread. The high heels concentrate the force into a smaller area than the wide, flat slippers, resulting in a much sharper pain for you. Similarly, the corners of a box-shaped tunnel will feel a different force from the walls. Given that they've been digging tunnels since the seventeenth century, engineers have a deep understanding of how to manage this difference, to ensure the tunnel remains structurally sound for many, many years.

---

* Later, there was a tragic collapse of one section of the tunnel, which killed a driver. This was not a result of the tunnel design – rather, it was down to an inexcusable string of communication failures that led to the wrong adhesive being used in the ceiling supports.

Bore

As evidenced by the nickname given to London Underground, boxes aren't the only way to go – tube-shaped tunnels are pretty popular, and for that, you'll need to look to deep bore tunnelling. How do you bore a tunnel? Talk about trains! *Yes, I fully stand by that joke, thank you very much*... One of the benefits of bore tunnelling is that, apart from where the machine is put underground and taken out again, the tunnelling process happens entirely below the surface. This minimises disruption; a huge bonus in densely populated urban areas. Another benefit of bored tunnels is that they have a circular cross-section, so all of the forces acting on it from the surrounding rock are evenly distributed – there are no flat surfaces that could bend under load.

To dig such a tunnel, you bring in a Tunnel Boring Machine (TBM), usually nicknamed a mole. These cylinder-shaped machines can munch their way through almost any rock type, and they are huge – the world's largest TBM has a diameter of 17.5m (57.4ft, equivalent to four double-decker buses parked on top of one another).[*] To be honest, I don't think that a mole is the best animal equivalent to a TBM – I prefer to think of them as earthworms. Worms eat, push forward and expel whatever is left over, and while there are lots of different types of TBM, they pretty much all do those same three things.

At the front, they have a circular face covered in incredibly hard teeth made from tungsten carbide. The cutter-head rotates, breaking up the rock in front of it. This excavated material is swallowed through an opening in the face (some would call it a mouth) and it is carried inside the body of the TBM using a rotating conveyor belt. There, it is mixed with various additives (rather like saliva or stomach acid in some animals) that turn the rock into something with the consistency of toothpaste. After digestion, this goo is expelled out of the back of the TBM, and it travels along a conveyor belt

---

[*] Nicknamed 'Bertha', this behemoth spent two years stuck under Seattle. As of January 2016, she was finally on the move again.

until it reaches a processing facility above ground. There, the goo is filtered and treated, with much of it reused in other building projects. Because of their shape, TBMs produce smooth tunnel walls, significantly reducing the cost of lining the tunnel. Far behind the cutter-head of the most advanced TBMs, large robotic suction arms called erectors pick up curved segments of concrete and place them along the new tunnel walls to form a complete ring that supports the tunnel. As the TBM moves forward, more and more of these rings are put into place, until the tunnel is fully clad.

All of this I was lucky enough to learn from the team at Crossrail. Somehow I managed to convince them to let me walk through TBM tunnels at both Woolwich Arsenal and Tottenham Court Road stations, and I have to say, I was flabbergasted by the scale of the project. To be completely honest, getting the chance to hang around in high-vis clothes, and talk to civil engineers about awesome infrastructure projects, was the reason I wanted to write this book.

Anyway, the Crossrail tunnels are 6.2m (20.3ft) in diameter (almost twice as wide as those on the Tube network) and they were dug by eight specially designed TBMs, each weighing in at 1,000 tonnes. These tunnels are, in effect, a huge, tube-shaped jigsaw, with over 200,000 concrete segments used to clad them. And TBMs can do all this digging-and-cladding lark at the rather impressive rate of 100m (330ft) per week. It may not sound all that fast when compared to your car, but looking back into history provides some context. It took Brunel *16 years* to dig the original Thames Tunnel between Wapping and Rotherhithe (406m, 1,332ft in length). Crossrail's new Thames Tunnel between Plumstead and North Woolwich took just *eight months*. I knew that vast quantities of soil and rock had to be excavated to create tunnels, but until I spoke to engineer Steve Boyle, I had no way to picture it. On my visit, he gave me a factoid that made my jaw drop: 'At the height of the project's tunnelling phase, the equivalent of 120 to 140 trucks fully loaded with excavated material left the tunnels EVERY DAY.'

This is not to say that TBM tunnelling is without its challenges. Imagine you're playing Jenga (or any other generic block-stacking game) but instead of a small piece of wood, you're removing tonnes of rock and soil from underneath existing buildings. What effect do you think that might have? The TBMs support the structure of the tunnel until the concrete rings are put into place, but vibrations from the excavation can loosen the soil and rock around it. Unless this is corrected for, anything built on top of the tunnel could sink, causing irreparable damage. Tunnelling engineers use a process called compensation grouting to minimise the risk to buildings while displacing tonnes of excavated material.

During the planning phase of a project, all of the buildings that sit along the proposed tunnel route are identified and have lots of sensors added to them. These sensors are constantly monitored and if any building shifts by even the tiniest amount during construction (we're talking thousandths of a millimetre here), an alarm is sounded at the control centre. And this is when the grouting team, the unsung heroes of tunnelling, come in. They work in vertical tunnels a few metres wide called grout shafts. From the shaft, small-diameter pipes are installed that radiate out into the surrounding ground. Through these pipes, engineers can pump a cement-like substance to exactly where it needs to go – directly under the structure at risk of sinking. The grout firms up the area where movement is predicted, to compensate for the loss of soil, and any sinking that might otherwise happen. It's like a filler for the wrinkled face of a city. This technique has been used in urban tunnelling projects all over the world, saving countless historic buildings that might otherwise have been lost in the name of progress. I am a fan.

Tomorrow

Much as I love them, trains couldn't really be accused of being futuristic. Rather, like so much of a city's great infrastructure, the best approaches have been those that gradually evolve over time. They keep up with, rather than leaping ahead of, what's available, but that's because we want this stuff to just work. Blue-sky science, with its promise of hover-cars and space elevators, is always sexy, but sexy doesn't get you anywhere if it breaks down all the time. Above all else, underground networks need to be reliable, robust and repairable. But, from speaking to rail systems experts from across the world, it seems that this attitude is slowly beginning to change. It is no longer enough to have a system that works reliably for a long time. Rail engineers are now actively on the hunt for novel ideas that break the mould, helping the whole system to become more efficient in the process.

Move

First is magnetic levitation (maglev) – OK, not entirely a future technology as it's already in use, but the next generation of maglev-inspired trains are the stuff of sci-fi dreams. If you've ever travelled around parts of Asia, you may already be familiar with maglev trains; commercial systems are in operation in Shanghai, China and Linimo, Japan. With a top speed approaching 600kph (373mph), they are the fastest trains in the world, and that's mainly due to one thing: they don't have to fight friction. Unlike standard trains, maglevs aren't in contact with anything. Their magnets let them float above the track.

Levitating something using magnets is a complex process. The idea is to counteract the gravitational force that would normally anchor an object to the ground with a magnetic force.* Importantly, to be considered truly maglev, magnets must also be used to propel and stop the train. So, how do they work? If you've ever played with a pair of magnets, you'll

---

* There is a long-standing myth that says that gravity is much weaker than electromagnetism. This is *only* true if you're as small as an atom. If you are a planet, gravity is the dominant force – electromagnetism doesn't even come into the equations that define planetary motion. Fundamentally, there is no way to directly compare the absolute strength of the two forces – depending on what it is you're looking at, the ratio between the forces changes.

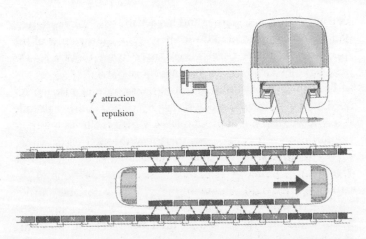

Figure 6.4 Some maglev trains act as a monorail (top image), with magnets mounted to the train body, and those on the rail just help with steering, but more advanced systems use alternating magnets to accelerate, slow down and stop the train (bottom image).

know that sometimes, no matter how hard you try to make them touch, they will repel each other. This is a primitive form of levitation, but it's very unstable – the magnets can slide sideways or flip over, which would be a problem if they were attached to a train.

To achieve both stability and speed, maglev systems use a combination of different magnets – some to levitate the train and others to propel it. Depending on which version of maglev you go for, these magnets are either attached to the underside of the train, which then straddles a steel rail (top image), or they're attached to both the train and a pair of rails called a guideway (bottom image). Either way, these trains are never in contact with the rails, and are propelled along on a cushion of air using a pattern of magnetic fields that repel and attract, pushing and pulling the train.

Because maglev trains float, they can reach speeds only dreamed of by traditional trains, but to do this, they need to literally pierce through air. The long noses on Japan's trains go some way to achieving that, but only up to a certain speed.

Beyond that, it gets harder and harder to 'beat' air resistance, and the power needed to do so increases with the cube of the speed. So, to travel **three** times faster (from 160kph to 480 kph) you need 27 (3 x 3 x 3) times more power!

I can hear you thinking, 'How do aeroplanes manage it?' Well, part of the reason is that the air at a plane's cruising altitude is less dense than it is at ground level, which results in lower air resistance, so planes can push their way through pretty easily. Closer to the ground, air molecules are a lot more tightly packed. Dense air = more air resistance = stronger vibrations in the structure of the train. This won't make for a comfortable journey for train passengers, and that's even before the inevitable motion sickness that we may feel at ultra-high speeds. The air-punching issue is true for all trains – hybrid, electric and maglev alike.

With some changes, maglev technology has the potential to go much further. Enter the Hyperloop – a tube which has had most of the air removed from it. With very, very few molecules left, the air resistance experienced by a capsule moving through it (accelerated using magnets) would be absolutely tiny. So with only a small amount of power, you could produce incredibly high speeds. Similar ideas for 'vac trains' have been around for decades, but, as seems to be the standard with anything involving Elon Musk's team, it finally has some real investment behind it, and tests of the system started in 2016. Definitely one to watch for future urban transport.

Another magnetic system that has caused great excitement in the media is the SkyTran – a public transport system based on pods that travel on a raised, magnetic levitation monorail. The idea is very similar to a maglev train except the rail would be suspended high above the street, and passengers would travel in small pods. The major reason that people are so excited about SkyTran is because it is being developed in collaboration with the Ames Research Center (part of the US Department of Energy). I must say, while I love the idea of it – effectively an elevated taxi rank run on magnets – I'm not convinced that it will be widely used in the cities of the future. The cost involved would be prohibitively high, so it would need to become popular extremely quickly for it to make financial sense in a city. But trials are in the works at a technology campus in Tel Aviv. Engineering and science-wise, there's not a lot wrong with the idea, but I think it still has a long way to go before it makes its mark. I would be very happy to be proved wrong though!

Even if these systems don't go beyond the 'incredible idea' stage (and frankly, I think the cost remains the only major barrier), we are starting to see other train technologies that are truly futuristic. Globally, the rail industry is looking for alternatives to carbon-heavy fuels – and electrification only gets us some of the way. Most trains still depend on fossil fuels to produce the electricity on which they run. We talked about hydrogen cars in Chapter 5, but for many rail experts, hydrogen-powered trains are gaining traction. Ultimately, fuel cells are on-board electricity generators, which release electrons when they react hydrogen with oxygen to form water. Given that the majority of urban rail systems are electrified, you can probably see how trains carrying their own source of electricity could catch on. Hydrail (see what they did there?) could offer all of the benefits of electrification, while removing some of the obstacles; there'd be no need for overhead wires or extra rails to carry the electricity. Huge research projects are ongoing and in 2015, two major hydrail pilots were launched – a new tram system for Qingdao in east China, and another around the Burj Khalifa in Dubai. It is still early days but, as with the automotive industry, there seems to be a lot of excitement around hydrogen.

Control

If that's how the trains of tomorrow will move, how will they be controlled? I'm happy to report that we are moving away from fixed trackside signals – yippee! The key to this is a huge step up in communication, both between the train and track, and between the trains themselves. In this new data-heavy system, the trains act as 'moving blocks'. Instead of depending on signals to move safely, they use on-board sensors to define their own 'safe zone'. They continuously measure the precise location, speed, direction and braking distance of the train, and communicate this to data receivers in the track. In turn, the track uses radio transmitters to send the data to other trains along the same stretch.

With lots of safety margins built in, this kind of system could enable the widespread use of driverless trains, at least in underground systems, but we should agree on what we mean by 'driverless'. Thankfully, there is an official 'Grade of Automation' classification system to help (yes, honestly). It tells us that Level 4 systems are capable of operating automatically at all times, including navigating, managing the doors, monitoring the track and responding to emergencies. Sounds cutting-edge, right? Well, at the last count, there were around 40 cities that boast a system capable of operating at Level 4, including Barcelona, Kuala Lumpur, Seattle and Taiwan. But most choose to have an attendant on board to manage some of the traditional tasks of a driver. Copenhagen definitely wins on this one. Its Metro is run by a fully automated computer system, which analyses terabytes of data in real time and makes decisions based on it. It uses a combination of fixed and moving blocks to manage traffic, but everything is computer-controlled. A team of just four people work at the city's Control Centre, acting as backup and monitoring the system. According to Siv Bhamra, Principal Vice President at Bechtel, '[For rail,] automated systems are the most safety-assured systems we have. And this fact will only be reinforced as our cities become megacities, and our train systems ever more complex. Running several different signalling regions (which is very common in cities) would be like speaking five different languages a hundred times a day. Without autonomy in our trains, this would be impossible.'

Autonomy offers a bigger challenge for intercity trains though, mostly because the speeds involved are much higher. However, here again there is a real momentum behind future-proofing the system, starting with the adoption of digital signals. Across the EU, all new intercity trains must now be compatible with the European Train Control System

(ETCS), which the BBC described as being 'more akin to modern air traffic control than the century-old technology it replaces'. In ETCS, the driver receives all track and station information direct to the 'dashboard'. While it is still at the early stages of implementation, it is hoped that this system will eventually replace the more traditional signals across Europe, improving efficiency and increasing capacity on the network. Much as I would love to tell you otherwise, having fully driverless trains won't remove the need to stop at signals – it's more that the signals of tomorrow will be built directly into the train. There'll still be stops and starts on your journeys, but with tons more data available, they should be far less frequent.

Bhamra and his team are working on one project to investigate just how much we could do with this data. Nicknamed train zero, it is effectively a computer version of the next generation of train – a train-sim for those gamers amongst you. Using a purpose-built lab, prototypes of every single component of the train's considerable communications system will be tested to their limits. By working in collaboration with lots of different electronics manufacturers, scientists and engineers, Bhamra is confident that what they'll produce will be world-leading: 'This train simulator marks a huge frontier for us – combining models with real-world integration of electronics has never been done on this scale in the rail sector. This is genuine research – we expect, and hope, to be surprised by some of our findings!'

So, will future cities be awash with fully driverless, self-guiding trains? Maybe. Trains are definitely getting smarter, but huge infrastructure projects like urban rail take decades to plan and implement, often because of their cost. Conservative as it might seem, we're likely to see only a gradual change in most cities.

Grip
As I teased you with earlier, there is even a future solution for 'leaves on the line' – *lasers*! A laser originally invented by British engineer Malcolm Higgins is now being commercialised by

Delft University of Technology.* Mounted just ahead of a
train's front wheels, the system produces pulses of highly
focused infrared light to rapidly increase the temperature of
objects in its path. It's possible to heat up the leafy mulch on a
rail to temperatures of more than 5,000°C (9,000°F), vaporising
it instantly. Even better than that, this particular wavelength of
laser light is reflected off the rails, and causes no damage
whatsoever. In fact, it dries the rails as it clears the leaves,
preventing rust. Initial results suggest that the laser can do all
this even while travelling at speeds of up to 80kph (50mph). It's
currently being tested, but there are still some unanswered
questions – Professor Rolf Dollevoet from Delft said: 'The
question is … for how long the rails will remain clean [after
being lasered]. We'll measure the remaining friction over time
during rain, drizzle, frost and snowfall.'

There are other ways that tracks can be used: for one, they
can feed electricity back to the grid. As we discovered earlier,
regenerative braking systems are already in widespread use,
but the US city of Philadelphia has taken them to a new level.
Instead of trains recovering energy for their own use, any
excess power is directed to a centralised bank of lithium–ion
batteries, for use across the city's rail network. Once stored,
this electricity can be sold to the main power grid, helping
the city to top up peak demand, and earning the train
company money in the process. Other cities have begun to
incorporate this sort of technology too. One trial in the
Dutch city of Apeldoorn aims to use the electricity generated
by braking trains to power the charging point for the city's
electric buses – a very neat idea!

Mathematicians at the Université Paris Diderot have found
a rather unusual way to improve the efficiency of their Metro,
and it comes down to balancing supply and demand. If a
metro system doesn't have a Philadelphia-sized bank of

---

* For those who can never have too much information, this laser is an
Nd:YAG, which stands for neodymium–doped yttrium aluminium
garnet, $Nd:Y_3Al_5O_{12}$.

batteries to store braking energy, it only benefits other trains
that are accelerating at the same time. This paper showed
that slightly altering the train timetable (to sync braking
trains with accelerating trains) could reduce the energy
consumption of a metro system by up to 10 per cent. Recent
work from the University of Birmingham used maths to
improve braking trains too. For an F1 driver, the key to a
successful race is how and when you brake or accelerate. The
team took the maths behind this idea and developed the
perfect braking routine that both slows the train down
sufficiently, and captures the largest amount of energy
possible from the motor-generator. For train services around
Japan, this kind of energy-saving is far more than a luxury.
In the wake of the Fukushima disaster in 2011, the Japanese
government shut down all of the country's nuclear reactors,
resulting in a hugely depleted power grid. By reducing
energy waste through improved regenerative braking, the
Japanese provided themselves with a back-up power supply
in their time of crisis.

Making better use of electricity that might otherwise be
wasted is part of a much larger trend. As we discussed in
Chapter 2, we're beginning to see the use of multiple energy
storage solutions, together with a gradual move away from an
all-encompassing central power grid. This will have a major
impact on the way our trains are powered.

Tunnel

Now, I'd like you to carefully pass a pencil through a gap.
The gap is just wider than the pencil – the difference between
them is no more than the thickness of a few toothbrush
bristles. You can't touch the walls of the gap, and you need to
keep your eyes closed. Could you do it? Now scale the whole
thing up by a thousand times, and add in a tube station full of
commuters, and you get some idea of the challenge that faced
Crossrail engineers. Upwards of 150,000 people use London's
Tottenham Court Road station every day, and all take the
long escalators down from street level to the platforms. But
unknown to many, on one normal weekday in 2014, a huge

digging challenge was in progress. Engineers guided a 7.1m
(23.2ft)-wide TBM through the gap between the escalator
and the platform tunnel. The monster machine passed just
80cm (31in) above the top of the tunnel and 35cm (14in)
under the escalator's foundations, and they did it without
closing the station.

The trick to this remarkable achievement was knowing
exactly where they were, and where they needed to go, at all
times. This is particularly tricky underground because the
go-to option of GPS simply doesn't work (discover why in
Chapter 7). Even if it did, GPS wouldn't be nearly accurate
enough to guide a tunnel-boring machine. Navigating
tunnels starts many years before a single trench is dug. The
detailed geotechnical survey defines the overall route, but
then surveyors 'walk the line' to measure the position of
thousands of points along it, noting each one's orientation
(e.g. 30° north) and the depth it will be below the surface. A
series of equations are used to join the dots into a single,
smooth curve. This gives us an accurate route map that the
TBM can follow.

But all of that is done above the ground. Underground, the
TBM needs to be guided along that route – otherwise it's
driving blind. Tunnel engineers use a system called laser-
based tracking to measure the constantly changing location of
a TBM. There are three key bits: a laser, a detector and a
control unit. The laser station (called a theodolite) is effectively
a movable telescope with a built-in laser, mounted onto the
tunnel wall. The detector is a fancy digital camera, and it's
installed on the back of the TBM. The control unit is the
brains of the operation – it stores all of the survey reference
data, and collects and analyses everything that comes in from
the theodolite and the sensor. Mick Lowe from ZED Tunnel
Guidance told me: 'The laser beam's position on the detector
can give us a surprising amount of information.' For example,
it tells us if the TBM is too far to the left or right of a reference
point. It can also tell us if the TBM is on the wrong angle,
or if the cutter head is off-centre. Using a radio link, the
laser and detector feed this information to the control unit,

which analyses it live, and displays it to the TBM driver. This up-to-the-second update means that the TBM can be kept on track, and it helps the driver to make any small adjustments as soon as they're needed.

As the TBM ploughs onwards, the theodolite will need to be carefully moved to a new position, so as not to lose sight of the detectors. When it has been moved, devices called retroreflectors are installed in the old location. Because they bounce light back in a single direction – parallel to the beam of light that originally hit it – they help surveyors to check the TBM's position relative to the planned route. And it seems that tunnellers are a rather exacting bunch. I was told about a recent tunnel project in France, where a pair of TBMs each started at the end of a route several kilometres long. When they met in the middle, they were out by 5mm (0.2in).* The entire team were inconsolable, and there were even a few tears shed. A difference this tiny falls well within the design margin for any tunnelling project, but they weren't happy.

Laser theodolites and other sensors are also used to constantly monitor movement in any surrounding infrastructure during tunnel construction. In the case of Tottenham Court Road, the existing tube tunnel moved by less than 3mm (0.1in), despite the 1,000-tonne TBM passing 80cm (31in) above it. According to rail Project Manager Linda Miller, 'Tens of thousands of motion sensors were installed across London years before we put a spade into the soil, and they will remain there for years more, continuously analysing the health of the structures.' Cities, by their very nature, are densely packed places, and without laser-based systems like these, we'd be hard-pressed to create tunnels at all. As today's cities grow into tomorrow's megacities, being able to 'thread the needle' through existing infrastructure will be ever more important. Next time you're walking around your city streets, look out

---

* This means that for every metre of tunnel the TBM dug, the navigation was out by much less than the thickness of a sheet of paper.

for those sensors on the corners of buildings. They'll tell you where the tunnels are.

I'm being terribly ageist though, because not all tunnels are new and shiny and built using huge 'worms' and lasers. We have to discuss the older ones too, because, really, they're not going anywhere! Most consist of a complex patchwork of brick, iron, steel and concrete, constructed and repaired using different building practices, over hundreds of years. So, it's very likely that even if you get hold of a tunnel's original plans, they won't in any way reflect its current state. Can you imagine, then, how difficult it would be to deconstruct such a tunnel, or how long it would take? In reality, it's completely uneconomical to replace ageing road and rail tunnels. So, if we can't take them out, we must maintain them.

Poke

'Rail tunnel inspection is still carried out by stick – it's basically a Victorian solution to monitoring Victorian structures.' Not quite the response you expect when you ask about the current state-of-the-art practice in all things tunnel-maintenance. But Dr Nick McCormick from NPL wasn't joking.

Tunnel examination in the UK and many other countries relies almost entirely on tactile and subjective visual inspection techniques. Individual inspectors walk through the tunnel in the dead of night, armed with sticks and torches. They then check for any suspicious-looking cracks or bulges in the fabric of the tunnel, tapping anything that seems troublesome with the stick. They log the position of issues that can't be easily fixed for further repair. Twelve months later, they walk through the same tunnel again, and look for any changes. There is absolutely no doubt that these inspectors are highly skilled and experienced, and that they are successful at identifying many pressing safety issues. But they're being asked to carry out an impossible task when it comes to identifying the tiny changes that are the precursors to faults.

*Figure 6.5 Behold – an actual stick used for tunnel inspections.
(Credit: Nick McCormick)*

In order to move away from this approach, McCormick, his NPL colleagues and two partner organisations developed the DIFCAM (digital imaging for condition asset management) system, specifically for examination of the interior of rail tunnels. Built on a road-rail vehicle (a 4x4 with additional guidewheels), it consists of 11 digital SLR cameras – 10 to image the tunnel and one that points down at the track – and four flashes to illuminate everything. On the front of the vehicle, there is a laser scanner that rotates 200 times a second. Its job is to continuously scan the tunnel, and gradually build up a map of its shape.

Each time the vehicle moves one metre, all 11 cameras take an image of the illuminated tunnel. When combined with the laser scans, the entire tunnel can be recreated digitally, with sub-millimetre resolution. By accurately noting the DIFCAM vehicle's location, it's possible to carry out the same measurement series several months (or years) later.* The new images can then be compared to the old, and any changes in the tunnel wall's appearance are automatically

---

* The NPL team designed its own odometer to do this. It is a device similar to that found in your car – it measures distance by knowing the circumference of the wheel. Despite its simplicity, it is accurate to 20mm (0.8in) in every 500m (1,640ft).

identified; for example, DIFCAM can tell you exactly how much a crack has grown in 12 months. The tunnel inspectors can then use their knowledge and judgement to make a decision on that defect. They can decide if it needs to be repaired, and if so, how soon. Far from taking the job away from a tunnel inspector, this system could provide them with more information, allowing them to apply their skills in the best possible way.

Another consideration in tunnels is the piston effect. I learned about this in my first week of living in London. As I emerged onto a busy platform at Clapham South, a huge gust of wind ripped off my scarf and lifted my dress clean above my head. I've never quite got over the embarrassment. So what causes it? Out in the open air, as trains move they easily push air molecules aside in all directions. But in tunnels, that air is confined, so it can't get out of the way quite so easily. This is especially true for train tunnels that are generally only slightly larger than the trains travelling through them. Just like a piston in a car engine, as a train moves through an underground tunnel, a wall of air molecules builds up in front of it, producing an area of warm, high-pressure air. Behind the train, you get a corresponding area of low pressure – so low that it can even act as a mini-vacuum, sucking the train back. Add to that the fact that many metro stations have areas above ground, or open out into the street, and you end up with a turbulent exchange of air. A wind that is unique to the underground, and one remarkably skilled at embarrassing people.

Although it sounds destructive, the piston effect can help ventilate train stations – in fact, it's often the only form of ventilation. However, we're seeing a growing number of so-called forced ventilation systems, based on a fan and a pump. These are very costly, in terms of both money and power consumption, but with growing worries over air quality in many major cities, they're often seen as the only option. So it was somewhat surprising that research from a group in Barcelona showed that, in larger stations, the air quality on the platform actually improves when forced ventilation is switched off. While this is not the case for all stations everywhere, this work has shown that with careful station design, energy-guzzling ventilation systems may well become a thing of the past.

Made

Future trains, tracks and tunnels are likely to be much more clever in their use of materials too. Way back in Chapter 1, we met concrete. But through my chats with various tunnel engineers, including Andy Alder, I learned a lot more about the grey stuff, including the fact that concrete-spraying robots exist. Controlled by a skilled expert wielding something that looks suspiciously like a game controller, these robots spray pre-mixed concrete directly onto a 'raw' tunnel wall. It rapidly gives structural support to a tunnel, and it has a secret superpower too. Mixed into the gloop are tiny fibres of glass and steel, each with an important role to play. The steel fibres act like mini-versions of rebar (the steel bars used in reinforced concrete). As they mix and disperse through the concrete, they form a random mesh, which adds mechanical strength. It also makes concrete more ductile, letting it bend and stretch without breaking.

The addition of glass is particularly interesting for tunnels, because here its job is to act as fireproofing. Glass fibre does not burn – when temperatures rise, all it can do is melt. Even approaching 600°C (1,110°F), it retains some of its strength, which could buy precious seconds in the event of an underground fire.* The nightmare of being caught up in such a tragedy may not bear thinking about, but that is exactly what tunnel designers must do. Ensuring that the tunnel remains structurally sound is at the very top of their list, and those tiny fibres of glass could make all the difference. As our cities get more tightly packed, we will continue to use more of the

---

* Sadly, there are still occasional tunnel tragedies – the fire in a Salzburg tunnel back in 2000 claimed the lives of 155 people, and in 2002 a fire in a rail tunnel on the outskirts of Cairo killed 383 people.

space under the ground. Materials like this spray concrete will be an enabler of the next generation of ultra-safe tunnels.

As you may remember, standard concrete is a mixture of aggregate (bits of rock) and Portland cement (a fine powder that binds the aggregate). Cement has long been the 'bad boy' of construction materials, and the industry's environmental impact is enormous. Some suggest that it accounts for 5 per cent of the world's CO_2 emissions. But now a British company has developed a concrete that is completely free from cement. Instead it uses a waste product from the steel industry (ground blast furnace slag) as the main binding agent. Many in the construction sector remain sceptical of the material – I was told that making concrete without cement was like 'baking a cake without flour'. But trials have shown that it performs as well as traditional concrete in many applications, and even requires less water to cure. So, we may yet see it everywhere. As a big fan of the flourless chocolate cake, I'm rather optimistic about the claims.

The trains of tomorrow are also going to be made of sterner stuff than today's fleet, which rely heavily on fibreglass to minimise weight. At the end of 2014, engineers at the Fraunhofer Institute in Germany unveiled a prototype train cab unlike anything we'd seen before. Made from a strong and lightweight metal foam, this new cab outperforms both fibreglass and sheet metal in crash tests. The foam is 25mm (1in) thick and is a composite of magnesium, silicon and copper, sandwiched between two aluminium plates. I should say that metal foams themselves aren't new; they were invented in 1986, but they've always been difficult and very expensive to make. These new composite foams are a whole new breed – they can be produced more reliably and at a lower cost than the original foams. At the time of writing, more advanced prototypes were being designed, so we can expect to see foamy metal cabs on European trains in no time.

From my conversations with the rail industry, it's become clear to me that, while it is still conservative, it is beginning to look further ahead, and it wants to become more innovative and more scientific in its approach. So although we're unlikely to see flying, supersonic trains any time soon, there are huge changes coming down the tracks (pun absolutely intended). Arguably, the biggest challenges facing transport aren't all so tangible, but to understand them, we need to start joining some of the dots we've laid out in *SATC*.

Connect

We are almost there. We have reached the heights to build a skyscraper, and donned our hard hats to dig a train tunnel. We understand where our electricity and water come from, and where our waste goes. We've taken to the open road in all manner of vehicles and enjoyed the view from some of my favourite bridges. And through all this, we've glanced into the future of cities, to understand how all of those systems will change the world tomorrow.

However, we're still missing a few less obvious elements: those networks that are either completely invisible to us, or so important that we simply take them for granted. We've yet to discuss the role that money and trade have played in shaping the urban landscape, or the challenge of feeding those who inhabit it. And of course, it's not just physical objects that flow through our cities – information can be shipped at the speed of light, along the ultimate superhighway.

It's a complex web to navigate, no doubt, but by doing so, we'll understand what ties urbanites to each other, and the connections that draw the city together

Today

At first glance, today's cities haven't changed all that much in the last 50 years. OK, they're a bit taller, sleeker and generally shinier, but fundamentally, they're still made from the same building blocks. If we dig a bit deeper though, we'll see that the city's DNA has been subtly but irreversibly altered, and it's mainly down to our love of data. Digital information has become the only truly international currency of the world's cities. But for many of our hidden networks, data could be seen as merely the modern icing on an ancient cake. Long before strings of 1s and 0s could be transmitted, information (and goods) made their way to us via a network of shipping routes that criss-crossed the planet. Yes, it was a pretty slow process, but shipping sowed the earliest seeds of globalisation, making the world a little bit smaller. All of that began in the city.

Ship

Forget the stock market. Throughout history, it's a city's port that has been its true economic hub. Humans have traded by ship for thousands of years, and many of the world's major cities, such as Shanghai and New York, still owe their development to the early success of their port.* These days, it's pretty unusual to find a port in the city centre (the high cost of land has a lot to do with that), but the link between the docklands and the city remains strong.

Talking about boats and ports might seem a bit medieval, but even today, almost all global trade (between 80 and 90 per cent, depending on who you ask) is carried by sea. Every single day, at ports across the globe, thousands of tonnes of imported goods are moved from massive ships onto trucks and trains, while exports go in the opposite direction. Most people prefer to take to the sky in order to travel, but goods are all about the cargo ship, and our shipping lanes have never been busier. As always, don't take my word for this stuff – next time you're online, have a play with the interactive maps found on the Marine Traffic website. There, you can not only lose hours of your life, but, in real time, track the position of every ship carrying a transponder. This system uses GPS (and yes, we're going to learn about that very soon) to determine a ship's position, course and speed. Along with details of the ship itself, this data is transmitted using radio waves to all ships in the surrounding area, without any input from the crew. It acts as a kind of automated beacon to say 'I'm here, I'm <this big> and I'm heading your way'. All large cargo ships and passenger ferries carry these transponders, and although they were originally

---

* Exactly how long humans have been a sea-faring species remains a matter of debate. And as I am not an anthropologist, it is a debate I have chosen not to enter.

developed for safety reasons, the data has now gone global. It can be accessed via the snappily titled 'Automatic Identification System', and from there, the Marine Traffic people use it to produce gorgeous density maps of ocean traffic.

At the end of 2014, Jean Tournadre, an oceanographer at the French Institute for Marine Research, published a paper that took shipping information to an entirely new level. He used satellite data collected over a 20-year period, and showed that there were four times as many ships at sea in 2012 as there had been in 1992, and mostly in the world's major shipping lanes. The cool thing is that the data Tournadre used, called satellite altimetry, wasn't designed for this purpose at all; it usually measures the height of the sea. Satellite altimeters do this by sending a pulse of microwaves from a satellite to the ocean surface, and waiting for its reflection. By precisely measuring the time in-between the pulse leaving and returning, it can determine the distance between the satellite and the ocean surface. And in the same way that an orange flung at a wall will be a bit deformed by the process, a pulse of microwaves also changes shape when it hits the water. This deformation tells us if the sea is rough or smooth, and it can be used to image ocean currents.

But Tournadre realised that the same data could also be used to pinpoint reflective objects that stick out of the sea, *i.e.* cargo ships. By looking at a specific microwave wavelength from seven different satellites, he could measure height differences between the ocean surface and the deck of a cargo ship. From space! This is equivalent to measuring the length of a 30cm (12in) ruler from a distance of around 20km (12 miles).* His analysis showed that excluding the years 2008–9 (economic crisis anyone?), global maritime traffic has grown by more than 10 per cent per year since 2002, with large cargo ships accounting for much of that growth.

---

* I've made some simplifications here. First, I took the orbit height of Envisat, one of the satellites Tournadre used (790km, 491 miles), and I'm assuming that a loaded cargo ship sits 10–12m (33–39ft) above the ocean surface (I asked around, and this seemed reasonable).

It is these cargo ships that populate our urban ports, carrying the products that fill shop shelves. Regardless of where they come from, they all have one thing in common: an object that is always exactly 2.44m (8ft) wide – the shipping container. There is a subtle genius in these humble steel boxes. Having a standard width means that they are compatible with various modes of transport, so cargo goes seamlessly from ship to rail and truck. They can all be moved using the same type of crane, and vitally, can be stacked and bolted together, making the most of every centimetre of space on a cargo ship. More recently, half-length containers have become popular, as have slightly taller ones, but the width of these durable boxes, made from corrugated, painted steel, remains constant.* In his excellent book, *The Box*, author Marc Levinson argues that it was their widespread adoption that slashed the cost of transporting goods, effectively creating the massive shipping industry that we all rely on today.

Some cargo ships are humongous (yes that is a real word, thank you). The world's biggest is the Maersk Triple-E. It is 400m (1,312ft) long – that's the length of four rugby pitches – 59m (194ft) wide and 73m (239ft) tall. Each Triple-E can carry 18,000 standard 6.1m (20ft)-long containers. If those containers were placed end to end, they'd stretch all the way between the famous university towns of Oxford and Cambridge – a distance of almost 110km (68 miles). And how many people look after this behemoth? It has a crew of just 20. This is because much of today's shipping process is automated, and that trend looks set to continue, as we'll discuss later.

Chill

Have a think about what you've eaten so far today. How much of it was sourced locally? Perhaps you started with a cup of tea (India) or coffee (Brazil), maybe with a bowl of nutty

---

* While I hate imperial units, they are the easiest way to describe the dimensions of a standard shipping container: 20ft long x 8ft wide x 8.6ft tall (or 6.1m x 2.44m x 2.62m). Others can be 10 or 40 feet (roughly 3m or 12m) long and the tallest containers are 9ft 6in (2.9m).

muesli (various). Then maybe a salad packed with tomatoes (Spain) and avocado (Mexico), or if you're more of a carbohydrate fan, maybe you had rice (China) or bread (UK) for lunch. Don't forget your mid-afternoon kiwi fruit (New Zealand), banana (Costa Rica) or chocolate bar (Ecuador). I'm not *trying* to make you feel guilty. I just want to point out how accustomed many of us have become to eating food from all over planet Earth.

It's true all year round too. If I need to find a yellow pepper in December in London, I don't go to a specialist importer – I just head to my local supermarket. It's not all that long ago that we could only consume food that was both local and in season, but for many people in cities, food seasons have lost all meaning. There is no international consensus on what qualifies as 'local' either. For many of us, the food on our plate travelled a long way to get there. Before we talk about the implications of this, or the science behind food production, let's find out how it's even possible for an orange from São Paulo to arrive fresh and ready to eat in Berlin.

It's all down to a concept called the cold chain (which I feel would be an excellent name for an ice-cream shop). All food degrades with time – once you harvest a fruit, you remove it from its own source of food (the tree or plant) and kick-start a series of chemical reactions that will eventually end in the death of the fruit. So, the key to long-distance travel for fresh produce is to slow down these reactions as much as possible. Keeping stuff cold is one way to do it, but other factors include moisture, light and air. The aim of a food exporter, then, is to minimise their produce's contact with all of these elements. Microorganisms in the air can land on the food, and fuelled by moisture, break its chemical bonds and gradually consume it. Reducing the temperature slows down the microorganism all-you-can-eat banquet (by deactivating the enzymes), which is why we keep so much of our food in the fridge. In certain produce, light can cause the vitamins and fats in the outer layers to degrade, changing its flavour and its appearance. This is why items like crisps and beer are stored in opaque or dark containers.

The idea of keeping foods cold and covered isn't a new one – ice-pits were first perfected by the ancient Persians, and by the 1800s, cold stores for fish, meats and dairy products were in widespread use.* Arguably, the first truly successful refrigerated ship was the *Dunedin*, which left Port Chalmers, New Zealand in 1882 packed with tonnes of animal carcasses and butter (yummy!). It arrived 98 days later in London with its cargo still frozen. Perhaps a little counter-intuitively, the *Dunedin* used a coal-powered system to cool down the food. It worked by squeezing air into sealed vessels and then releasing it into the hold; this sudden change of pressure caused the air to cool down.

If you've ever pumped up a bike tyre, you'll know that it can feel like hard work. That's because we transfer some of our energy to the air molecules being forced into the tyre. As we pump, the air molecules get squeezed closer together, increasing the rate at which they jiggle. Because temperature is really just a measure of how jiggly molecules are, squeezing air increases its temperature, making it hotter. If we then suddenly detach the pump from the tyre so the air rushes out, that stored heat energy is rapidly spread over a very wide area, making the surrounding air cooler. The same effect happens every time you spray deodorant, which explains why it feels cold.

Anyway, since the *Dunedin*, refrigerated shipping has become much more high-tech, and today's refrigerated containers (nicknamed reefers) are temperature-controlled masterpieces of engineering and technology. Reefers are a key link in the unbroken cold chain our food supply depends on. From harvest or production through to consumption, fresh produce can be kept at a constant, low temperature. What that temperature is varies according to what we're shipping. Oranges are shipped at the temperature range known, rather confusingly, as banana, which is 12–14°C (53–57°F). Other food products such as meat and seafood are stored at significantly lower temperatures

---

* For a brilliant read on the history of refrigeration, I recommend getting your hands on Tom Jackson's book *Chilled*.

(between -30°C and -16°C, -22°F and +3°F) and vegetables and fresh meat quite literally chill (2–4°C, 36–39°F). Reefers act like scaled-up fridge-freezers, and are plugged into the ship's power supply, or that of the train or truck that takes them to their final home. All the while, sensors measure the temperature, humidity and air quality inside the reefer. The data is sent via satellite link back to the exporter, so that they can be sure their cargo is in peak condition.

For more fragile cargo, such as lettuce and berries, a faster route is needed, which is where air travel comes in. The most exported air commodity in terms of tonnage from LA International Airport is food: vegetables, fruit and nuts represent more than 15 per cent of the total weight of cargo shipped through the airport. Frankfurt Airport even has its own Perishable Centre – a rugby pitch-sized shrine to the temperature control of fresh goods. But once the goods are in the air, the temperature control is pretty old school. Instead of relying on a computerised power supply, the cargo is stored in an insulated box containing a chamber of frozen carbon dioxide (also known as dry ice). It's a bit of a strange material – instead of turning from solid to liquid to gas with increasing temperature, it goes directly from solid to gas, in a process called sublimation. And because it does this at -78.5°C (-109.3°F), it can be used to keep things cold. A block of frozen CO_2 takes many hours to sublimate; more than long enough to ship a box of juicy Californian raspberries from LA to Moscow.

I promise I'm not evangelical about refrigerated shipping. Whilst I view it as a remarkable engineering achievement that helped connect the world, it clearly has its downsides. Now that I know a bit more about shipping, I still find it mind-boggling that my all-year-round addiction to oranges can garner profits for those along the supply chain, while still costing me relatively little. Yet, there is no doubt that environmentally, it would be much better to source the things we eat and drink locally. Moving food vast distances in fossil-fuelled container ships, planes, trains and trucks is more than a little wasteful. As our cities continue to get denser, the challenge of getting fresh food to urban dwellers will deepen. And refrigerated containers just aren't going to be enough.

Eat

One major thing we need to consider for today's cities is food production. Agriculture and urbanisation have long had a complex relationship. Many of today's largest cities developed precisely because that particular location offered the most fertile soils and the cleanest water, but because of that, lots of original farmland now contains houses. While this has effectively pushed productive farms further from urban centres, they remain inextricably linked. Food security is no longer viewed as a rural issue – with more people living in our cities than ever, it has shot right to the top of the urban agenda.

We know that a sustainable food supply doesn't come for free. As you might remember from Chapter 3, the water cost for different foods varies widely. In addition, there is also a carbon cost to the production of food. In 2013, the UN Food and Agriculture Organization (FAO) released a report that specifically looked at greenhouse gas emissions involved in livestock farming. Beef came out at the top of that list – every kilogram of protein produced the equivalent of 300kg of CO_2. Further down the list came cow's milk, chicken and pork, which each produce less than 100kg of CO_2 per 1kg of edible protein.* It's not just livestock that emit greenhouse gases either – the growing use of synthetic fertilisers has caused a rapid increase in emissions in crop farming too. The FAO have also shown that, in just one year (2011), the world's fertilisers produced the same amount of CO_2 as they had in the previous decade. Put simply, in order to keep up with

---

* This study looked at the entire production chain, starting from land use, through to processing and transport. The metric of 'CO_2-equivalent emission' was introduced by the Intergovernmental Panel on Climate Change (IPCC) in 2007 – it is the internationally recognised standard way to compare emissions from a range of greenhouse gases.

food demand, we're using significantly more fertiliser than we ever have before.

The other question is, are we urban dwellers eating the right kind of food? We're now used to seeing nutritional information on food labels, including calorie count and salt content, so surely we're making better choices? It's not that simple. According to the International Institute for Environment and Development, 'hundreds of millions of urban dwellers suffer under-nutrition'. The poorest city-dwellers still have to make the heart-breaking choice between buying food to simply fill their stomachs, and buying food that will nourish them. Energy-dense foods (like doughnuts) offer lots of calories but very little in the way of nutrients. However, they tend to be considerably cheaper than the nutrient-dense foods (such as fresh meats or vegetables) associated with good health. In too many cities to mention, maps of poverty overlap with maps of poor health and obesity, and that is not a coincidence. But on this I was way out of my comfort zone, so I called in some help from urban food security expert Dr Louise Manning, at Harper Adams University. She said, 'There are enough calories produced on the planet right now – but they are not produced where the cities are, and are not spread evenly.' For Manning, this leads to the fundamental question for the coming generation, 'How can you feed an urban population when the demand of megacities greatly exceeds what the local land around them can supply, both in terms of calories and nutrients?' The reality is that feeding the city is about more than just environmental cost, land availability or food volume; it's also about the nutritional quality and variety of that food.

When it comes to variety, there's one name that crops up again and again: botanist Nikolai Vavilov, possibly the most important scientist you've never heard of. In a way, it was the rediscovery of obscure research by an Austrian scholar called Gregor Mendel that shaped Vavilov's career. In work now considered the foundation of modern genetics, Mendel showed that certain desirable qualities in a plant could be deliberately passed on to future generations. Vavilov was among the first to apply Mendel's idea of genetic inheritance to crop growth – he realised that by cross-breeding certain varieties, you could produce crops that were hardier, more productive and longer-lived. Since the beginnings of agriculture, farmers have re-sown seeds from only the best crops, a very primitive form of genetic modification (and we'll come back to this). Because there was little understanding as to why certain crops outperformed others, it was seen as an art form. Vavilov and his contemporaries made it a science, and it's largely thanks to them that we have maintained the agricultural diversity we have today.

In two decades, Vavilov and his team collected more than 250,000 crop samples from all over the world, but these scientists did more than that. During the 900-day Siege of Leningrad, they took turns protecting the seed bank and, remarkably, nine of them chose to die of starvation rather than eat the precious seeds that they guarded. Their legacy can now be found on a Norwegian island, 1,300km (800 miles) from the North Pole. The Svalbard Global Seed Vault is effectively the world's food backup system. It can store up to 4.5 million varieties of food crops, and the thick layer of permafrost that surrounds the facility keeps the seeds safely frozen, even without power.* Why is this important? I'll hand over to the Svalbard team themselves: '[It is] the ultimate insurance policy for the world's food supply, offering options for future generations to overcome the challenges of climate change and population growth.' Seed banks like this not only provide us with a way to maintain crop diversity, but in times of disease or disaster, they can allow local farmers to restart agriculture.

Many of the diverse foods we eat today originated elsewhere. My homeland of Ireland has a strong association with potatoes, but they first grew in South America. Tomatoes may be a staple of the Mediterranean diet, but their genetic birthplace is

---

* The Norwegian government paid for the construction of the seed bank, and its operating costs are paid for by international governments, charities, foundations and farming organisations.

Mexico. And while New Zealand lamb is world-famous, sheep were actually first domesticated in Kazakhstan. This diversity has come from a long history of trade between countries, but so far, we've only talked about the food and the logistics of moving it. We're missing the one key element that enables all this to happen – money.

Stock

Ayn Rand described money as 'a tool of exchange' because it came about as a way of standardising trade, and of assigning value to objects, services and time. And in that guise, money became a vital link in the chain that connects cities to each other. In the modern world, money is, in itself, a tradable commodity, despite it having no intrinsic value.* I'm not saying money's not useful – of course it is, but many people struggle to understand the concept of value when it's discussed in terms like *securities*, *options* and *stock*. At the risk of sounding like a dinosaur, it used to be much simpler. Once upon a time, value was measured against the gold standard. Gold's value comes from the fact that it never rusts and it can be used in countless ways – from crafting a piece of jewellery to manufacturing a tiny probe for an atomic force microscope. In contrast, the value of a banknote is based on the number we've printed onto it, not on the value of the material itself. These days, financial trading drives money between major cities, and there's a lot of science and technology behind it.

I asked some friends what they picture when they hear the words 'Wall Street'. The consensus was, stress, swagger and slicked-back hair. But it seems that these days, New York's financial heart beats with a new rhythm. The quants are in town and they've come armed with PhDs in maths, physics and computer science. A quantitative analyst, or 'quant', is someone who uses mathematical methods to assess and predict

---

* Arguably, coins have *some* intrinsic value because they're made from metals with other uses. In 2015, the Central Bank of Ireland announced the phasing out of their smallest coins, because they cost more to make than they're worth (1c coins cost 1.65c to produce).

financial risk. Their approach echoes that of all scientists through the ages – they observe and measure data, in order to understand the bigger picture. The difference of course is that the financial market isn't a tightly controlled experiment. It changes in response to any number of things – a rumour that a CEO is standing down, the announcement of a new product or an escalation of violence in a war-torn nation.

Financial markets are turbulent, noisy places, so what quants try to do is to see through that noise. Increasingly powerful computers can help pick out previous trends in the data. And then, mathematical models designed by quants are used to suggest possible future trends.* Sounds a bit like gambling, right? It's more like playing with a loaded dice. No mathematical model can perfectly predict what will happen all the time (anyone who tells you otherwise is a joker), but a good one can give a measure of how likely a particular outcome is. For financial markets, the decision to buy or sell comes down to how big that probability is: if a model says there is a 58 per cent chance of the stock increasing in value, it might be worth the risk. A model is no more than a tool – just like a drill or chisel, judgement and skill are still needed to get the most out of it. And there is always risk involved, even when the maths is solid and the computers are world-class.

Some of these mathematical tools stem from rather surprising places. The year 1905 was a pretty good one for Albert Einstein – as well as publishing his Theory of Special Relativity (changing the way we view space and time), and explaining how light can generate electricity (the basis of all solar power), he also proved that atoms and molecules were real. Einstein's description of why particles suspended in a liquid or gas jiggle constantly showed that atoms must physically exist – before, they'd been just a useful mathematical trick. Much later, this work led economists to develop a model that described the continuously changing prices seen

---

* A mathematical model is nothing more than a series of equations that aim to describe the data we're seeing.

in financial markets. Related to this is the heat equation; physicists use it to describe how heat moves (and temperatures change) over time. In finance it is called the Black–Scholes–Merton equation, and it is used to predict future prices. From plasma physics, economists have borrowed the Fokker–Planck equation to better understand the distribution of wealth and populations… the list is endless. Maths and the economy go hand in hand.

There is another side to this too. High-frequency trading (HFT) has regularly hit the headlines over the past few years, and rarely for good reasons. It relies on powerful computers to make trades at incredibly high speeds – much faster than a human can think. So here, computers and automated trades are king. Financial markets have moved quickly since computers first came into use, but now trades are measured in microseconds – that is, 0.000001 of a second. These time intervals are so short that even the length of the fibre-optic cables between computers can have an influence!* HFT has become particularly popular in New York and London, and despite being accused of destabilising an already complex market, I'm not sure it's going to disappear any time soon. For a system this time-sensitive, we'll need the best clock available.

Time
With time-zones and calendars varying across the world's cities, you'd be forgiven for thinking that time is arbitrary, but it is another vital connector in the network of the city. Culturally, time is an obsession: we take it, we find it, we share it, we make it. Three of the most common nouns in the English language are even related to it – in the #1 position of common nouns is **time** itself, in #3 is **year** and #5 is **day**. And the logistics of city living are entirely bound to time, from the definition of the working day to the time stamp on your supermarket receipt.

---

* Nothing travels faster than light, but it still takes time to get from one place to another – the shorter the path, the quicker it arrives.

Given how crucial time is to our cities, how do we measure it? According to Dr Patrick Gill from NPL, we need three things: 'First of all we need something that ticks, next we need to be able to count those ticks and then we have to know how many of those ticks occur in one second.' In a typical wristwatch, the ticker is a piece of quartz (SiO_2). When given a small amount of power from a battery it vibrates (or ticks) approximately 32,768 times each second.[*] Quartz clocks do a pretty good job of keeping time, losing or gaining just a few seconds over the course of a year. But there is a whole layer of technology that relies on timing, and for these tasks, the quartz clock was never going to cut it. In 1955, two scientists at NPL became the first true 'time-lords' – their atomic clock would lose 1 second in 300 years, making it hundreds of times more accurate than quartz.

Cast your minds way back to Chapter 2 and the picture of the atom, with electrons whizzing around a nucleus. Electrons always sit in particular energy levels – think of them as the steps of a staircase. Each time the atom absorbs energy (say from a laser), its electrons can leap up to the next step. When they fall down to a lower step, they emit energy, normally in the form of light. This 'step-jumping' process is called a transition, and it is extremely picky. Exactly the same amount of energy is emitted or absorbed every single time an electron transitions from one step to another, and the bigger the jump, the more energy is exchanged. Because of this, if we measure the frequency of that energy (effectively, counting the ticks), we can use it to tell the time. It's like the atomic version of counting the swings of an old pendulum clock, only 100,000 times more accurate.

NPL's latest clock uses atoms of caesium. Inside its carefully controlled chamber, the caesium atoms are cooled down so much that they hardly jiggle at all. Then they are bombarded with microwaves to cause the outermost electron to leap between two specific energy levels. This is repeated

---

[*] This is the ideal, but the actual number depends on the size and shape of the crystal, and how it was processed.

over and over to find the frequency of the atom. Unlike quartz crystals, all electrons are identical, which means that this frequency never varies, making it a very reliable way to measure time. And, although you might not know it, this is at the basis of every measurement of time on (and off) Earth – every 9,192,631,770 times this transition happens, one second will have passed. Scientists at NPL and around 20 other labs, including the National Institute of Standards and Technology in the US, continue to work on increasingly accurate atomic clocks. There's a lot of jostling between labs as to who currently has the most accurate clock, but they're all honing in on systems that lose no more than one second in the age of the universe.

'What does counting a jumping electron have to do with city living?' I hear you cry. Well, for a start, financial systems and databases are at their best when accurate time-stamps are used, and as trading gets faster, the clocks need to get better – financial centres now all tap into atomic time. Power and electricity distribution is also reliant on the measurement of time; according to Rich Hunt from GE Digital Energy, 'When it comes to the modern grid, timing is everything. Without it, relays would trip and power lines could go out.' And last but by no means least, the internet and everything connected to it runs on Coordinated Universal Time (UTC), and that is based on a network of atomic clocks here on Earth. There are many in orbit too – every single Global Positioning System (GPS) satellite has an atomic clock on-board. Without time, we couldn't navigate around our cities.

To make GPS global, we need 24 satellites. Their orbits are designed so that, at any point on Earth at any time, at least four satellites are visible.* But GPS receivers (like the one in your smartphone) don't just 'know' where they are – they have to figure it out. Each satellite transmits information using pulses of radio waves, and because we know that each pulse travels at the speed of light (almost 300,000km/sec or

---

* GPS is owned and maintained by the US military. Other positioning systems do exist, but I'm just looking at one for now.

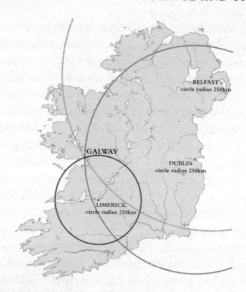

Figure 7.1 Trilateration in action, projected onto a 2D map. Each circle represents the 'view' of a single satellite. Using three, we can pinpoint our location precisely.

186,000 miles/sec), if we time it precisely, we can measure distance (distance = speed x time). In order to pinpoint your location, satellites use a process called trilateration, which we'll need a map to explain.

Say you found yourself lost in Ireland. You ask for directions but all you get is, 'Well, you're 187 kilometres from Dublin'. This is the same as having one satellite, and while that's helpful, it doesn't give you a direction to go in. So you ask another person, who says you're 250 kilometres from Belfast. Now you have two points of reference. A third person tells you that you're 74 kilometres from Limerick, and you have it – you're in Galway!* GPS satellites do the same thing, but instead of drawing circles as I have on my map, they draw spheres in

---

* Vague directions like this aren't all that implausible. This is how I was once directed to a hotel in Ireland: 'Turn left at the next crossing. Then stay on that road, and if you see the old church, it means you've gone too far, so come back a bit and look out for the house with a green, or maybe brown, fence. Turn right there and then I think it's the next left.' We made it in one piece. And the fence was grey.

space. Technically, three satellites is sufficient to define a location, but to get to the level of accuracy we need, we use four. Because timing is so vital to how it works, without atomic clocks, our GPS infrastructure would fall apart.

GPS relies on some seriously hard-core physics to work. First, Einstein told us that clocks in orbit run slower than clocks on Earth (Special Relativity) but that when we measure them from the Earth, they *appear* to run faster (General Relativity). And second, because they depend on time measured using jumping electrons, GPS satellites also have to include the (occasionally weird) laws of physics that rule atoms. Every single time you use your GPS, all of these effects have been taken into account. Which leads me on to another question I've been asked: **Why does my dashboard GPS struggle to find itself when in a multi-storey car park or tunnel?** The short answer is that GPS signals are rather weak by the time they reach your receiver, and so they don't penetrate solid objects very well. So trying to use a GPS indoors can be rather frustrating. Once you're out in the open air, the satellites will eventually find you. However, if you're driving around an area filled with skyscrapers, you might find the GPS signal will struggle there too – there's a reason cities are referred to as 'urban canyons'. But, in general, I'd say cut your GPS some slack. It has a lot of work to do!

Chat

In recent years, the way we communicate has changed beyond all recognition, and just as with navigation, satellites are at the heart of it. As of mid-2015, more than 1,300 active (data-transmitting) satellites were orbiting the Earth, and about half of those were communications satellites.[*] They receive information from various transmitters on the ground. The satellite then amplifies that signal and transmits it back to

---

[*] The Union of Concerned Scientists keeps an up-to-date list of all active satellites in orbit. Spy satellites are probably not included though. Nor are the 'passive' satellites that act as mirrors, to extend the range of the signal.

receivers all over the Earth. This is the way we transmit digital TV, telephone and radio signals, and by bouncing them off a huge network of satellites, we make worldwide coverage the norm.

Imagine you're broadcasting a TV show (in the form of a Bat-signal to the streets of Gotham) to a moving satellite. You'd need to continuously move the Bat-signal in order to keep up, but once the satellite disappears over the horizon, it's all over. Not ideal for those trying to watch a must-see TV show. It would be much easier if the satellite could stay in one place in the sky, so information could be streamed to it constantly. Croatian rocket engineer Herman Potočnik first proposed this very particular type of orbit, now called a geostationary orbit, in 1928, and it was later suggested as a communications tool by mathematician and sci-fi writer Arthur C. Clarke.

Orbits are a bit weird (and I say this as a former space scientist). If you throw a ball across a field, it will quickly lose its battle with gravity and hit the ground. If you could fire that ball out of some sort of cannon, it would be able to travel much further before hitting the ground. If we keep upping the speed (and ignore inconvenient things like air resistance or buildings), we could make that ball travel many kilometres along the ground before it fell to Earth. Satellite orbits are based on that same idea – once we escape the Earth's atmosphere, we no longer have to worry about air resistance or skyscraper-shaped obstacles. To get us there, we'll have used a rocket that can propel our ball (now called a satellite) to incredible speeds. It's not that the satellite isn't falling, it's just that its path follows the curve of the Earth so exactly that it never quite manages to hit the ground. With some careful calculations, it's possible to put a satellite into an orbit so that it moves at the same speed as the Earth rotates on its axis. This would give us, effectively, a stationary communications satellite that transmitters on the ground can point directly at 24 hours a day.*

---

* There are other communications satellites that don't stay in one position and orbit a lot closer to the Earth's surface. Yes, they're more complicated to use, but they benefit from stronger signals.

So what exactly are these satellites transmitting? Well, part of the answer to this might explain the confusion between the terms Wifi, 3/4/5G and phone coverage. This is especially relevant in the city, where people switch between them continuously, almost without noticing. It comes down to light, also known as the electromagnetic spectrum. If this spectrum was a 30cm ruler (arranged from short wavelength to long), all of the light that we can see would be found between the 8 and 9cm markers on the ruler. Everything else along it would be completely invisible to our eyes, but we can detect it in other ways.

The microwaves and radio waves we use for communications would fall between the 15 and 23cm markings, which gives us lots of space to squeeze in everything from GPS to mobile internet signals. To avoid any interference between them, all communications technologies must use specific parts of this band, called frequencies.* The 'G's and calls from mobile phones rely on 'ultra-high frequency' waves (on the ruler, between 16 and 18cm), and they extend their range by using a network of towers to transmit and receive data to and from communications satellites. Wifi doesn't need a satellite – it uses small transmitters (routers) to transform the 'wired' internet into wireless, and broadcasts it over a short distance, say, within a coffee shop.

Together, these technologies really put the smart into smartphone, possibly the most ubiquitous technology in city life. There's plenty of debate about how our need to be connected at all times will affect society, but I'm afraid I'm not the right person to answer that.† What we can look to are the systems that make it possible. In his book, *Physics of the Future*, physicist Michio Kaku famously said, 'Today, your cell phone has more computer power than all of NASA back in 1969, when it placed two astronauts on the moon.' He really wasn't exaggerating – the central processing unit in a typical smartphone is able to

---

* The higher the frequency, the shorter the wavelength. If our imaginary ruler was arranged in order of wavelength, it would be a mirror image of the one I've described above.

† Although I recommend this excellent XKCD strip to help put it into context: www.xkcd.com/1601/

carry out millions of calculations every second, making it about ten times faster than NASA's Apollo-era mainframe computers. And the complex software program that monitored the health of the astronauts and the environment inside the spacecraft could fit on a 1GB memory key... 167 times. So what's changed? Size and cost. Electronics have become consistently smaller, better and cheaper since the late 1960s. Advances in microelectronics and materials science have given us high-capacity hard drives, ultra-thin laptops and tech-packed smartphones. The trend towards tiny, low-cost components shows no sign of stopping, and may even enable the 'Internet of Things'.

Before we get into that, we need to talk about cabled internet. With the widespread availability of Wifi in urban areas, you might think that cables are on their way out – but you'd be wrong. According to telecoms giant NEC, 99 per cent of all international data is still transmitted through cables, lots of which lie on the sea floor. Consisting of 885,000km (550,000 miles) of cables, this hidden subsea network could wrap around the equator 22 times! But it came from modest beginnings. On 16 August 1858, a message was sent along a new telegraph cable laid between Ireland and Newfoundland, and 17 hours and 40 minutes after leaving the Emerald Isle, the telegraph arrived at its destination. Although hardly what you'd call speedy, this comfortably beat messages sent by ship (and that was the only other option at the time).

The same data now takes a fraction of a second to travel that distance, thanks to the latest generation of submarine cables built on a heart of glass. Flexible and hair-thin optical fibres can transmit data incredibly quickly – at about 70 per cent of the speed of light.* They work on the idea that if light hits a transparent glass surface at just the right angle, instead of shining through it, it will bounce back in (the inner surface acts a bit like a mirror). If you design the fibre correctly,

---

* It takes about 8 minutes 20 seconds for the sun's light to reach us on Earth. In glass, light moves a bit slower (around 200,000,000 metres per second, or 124,300 miles per second). But it's hardly what you could call a slouch.

light is effectively trapped inside; all it can do is bounce along it. Any information added to those light waves – say, a video stream – is carried right along with them, allowing us to communicate seemingly instantaneously. Transmitting information by cable is still considerably cheaper, faster and more reliable than satellite-based communications. So even in the built-up, Wifi hotspot-enabled environs of the city, we still depend on the fibre-optic cables that snake under the paving stones. Thanks to them, our homes and businesses can get online in the blink of an eye.* For many urban dwellers, being connected is now seen as a vital part of daily life, and this has altered the way we interact with the city. Just how far our love of information will go remains to be seen, but it definitely gives us lots to discuss in our glance at the future.

Tomorrow

We all know that predicting the future is rarely a good idea, which is why throughout *SATC* I've focused only on urban-related things that are both scientifically sound and technologically plausible. But if any chapter is likely to leave me red-faced in a few years, it's this one. Bridges, tunnels, buildings, power and water have all obviously changed, but it has been a gradual evolution, with small improvements being made over a long period of time. But in the invisible streams of the city – its data, money, communications and logistics – the change has been both rapid and profound. While this is all terribly exciting, it also makes it impossible for me to identify those technologies that will be the true game-changers. I mean, there are the huge, obvious trends: our love of connectivity, combined with reliable, cheap electronics has produced what's been called the Information Generation. We are now used to getting access to anything we want, from anywhere, at any time. And we know that with a growing population and a changing climate, food supply and security

---

* It takes about 0.4 seconds to blink. In that time, light could travel 80,000km (50,000 miles) along a fibre-optic cable.

will face growing challenges in the years ahead.* Exactly how the science and technology behind these things will manifest themselves remains to be seen. A vast amount of remarkable research is being carried out right across these topics – far more than I have the space to introduce you to – so consider this a very brief snapshot.

Farm

Our hunger for data will certainly extend to the food we eat. From continuous monitoring of soil conditions to GPS-enabled weed mapping, precision farming is bringing data to the fields. Although still at a relatively early stage, this approach is already showing promise. In one project led by Professor Rajiv Khosla from Colorado State University, water content was monitored across rough, undulating wheat fields in northern India. This data was used to flatten precise areas of the field – a seemingly small intervention. But the next harvest produced 17 per cent more wheat with only half the amount of water. This sort of efficiency in food production is going to become ever more relevant.

It has been estimated that about a third of all the food produced for human consumption is lost or wasted every year. And with wasted food comes wasted resources, especially water, transport and electricity. We know that certain foods are more resource-intensive than others, and with a significant proportion of urban dwellers already under-nourished, we need to consider the type and quality of food we provide. So, as well as the challenge of producing enough food for projected urban population growth, we need to find clever ways to make the entire food chain more efficient.

---

* Despite what you might hear, among scientists there is no real debate around the reality of a changing climate. A survey of almost 12,000 peer-reviewed papers on the topic showed that '97.1 per cent [of them] endorsed the consensus position that humans are causing global warming'. The only debate is around what we can be or should be doing about it.

There is a lot of pseudo-science around food. While I might be partly protected from it by my physics background, I was very happy to find a guide in Dr John Kerr from Science and Advice for Scottish Agriculture. He summed up the reality of the situation: 'In order to feed everyone in 20 years' time, we will need all the help we can get. If we choose not to use all of the technologies at our disposal, then *all* available land will be needed for food production, wiping out the world's ecosystems.' One of the technologies very likely to play a role is the genetic modification (GM) of crops, but it is hoped that this latest generation will move on from the 'frankenfood' debate of the 1990s. First, some context: modifying plants for a specific purpose isn't new. In what may be my favourite ever sentence in an *SATC* interview, Dr Kerr said, 'We've been selectively breeding crops ever since man stopped running around after stuff, and stood still long enough to bake bread.' The difference is that now, crops don't have to be modified on the scale of a seasonal harvest; through altering the structure of the crop's DNA, scientists can develop robust crops from the bottom up. The main motivation behind this is efficiency – traditional plant breeding takes a very long time and produces a lot of waste.

The latest approach to genetic modification is called precisely directed mutation, because it could allow scientists to create just the 1 per cent of improved crops we need, rather than discarding the 99 per cent of stuff we don't (e.g. for a potato, this included everything that grows above ground). There are lots of different techniques involved, from silencing particular genes, to inserting the beneficial genes from one plant into another. I can assure you no-one is inserting chicken DNA into crops. Contrary to what some may think, science, and scientists, don't operate in a vacuum – the social, environmental and political implications of research are never far from their minds. And nowhere is this truer than around the question of GM crops. Science will not provide all of the answers here; it is part of a much larger context (as it should be). There is still a lot of debate around what a GM crop even is, and yes, there are still unanswered questions around the long-term implications of 'biotech-ing' our crops. For me, that is exactly why this research is so important and should continue – burying our heads in the sand is not going to work. If we can find a way to produce safe, pest-resistant crops that can grow with less fertiliser, or to develop low-cost rice with extra nutritional benefits, or even design crops that can withstand the ravages of a changing climate, shouldn't we do it?

Aside from GM, there are other agricultural technologies being developed that have particular relevance to the city of tomorrow, and all come under the heading of urban farming. In a previously empty part of a Fujitsu computer chip factory in Japan, a vast indoor farm can be found. Larger than a football pitch, it came into being after the 2011 earthquake ravaged the region's farming communities, causing food shortages across the country. In it, low-power LEDs are used to provide the light, but that's not the cleverest bit. All green plants use the energy in sunlight to convert carbon dioxide and water into glucose to help them grow (basic photosynthesis). But this process works faster at particular wavelengths of light – mostly red or blue – so by using LEDs that match these wavelengths, the whole 'turning sunlight into food' thing becomes more efficient. As well as light levels, air temperature, humidity and carbon dioxide content can all be carefully monitored and controlled.

Japan is not alone in this. In cities including London, Newark and Copenhagen, sun- and soil-free farms are beginning to change the face of local food production. These are often called vertical farms because plant beds can be stacked on top of one another, allowing many more plants to be grown than would be possible in a field. But don't panic, this technology isn't heralding the end of traditional outdoor farming. There are limitations as to what can be grown in these greenhouses; almost all of the leading projects focus on leafy greens.*

---

* In August 2015, the crew of the International Space Station harvested and ate a crop of lettuce that they'd cultivated on board. Growing vegetables indoors is so cool that even astronauts are doing it.

And yes, all of this runs on electricity, but so far that seems to be outweighed by the other benefits, especially for an urban consumer. They're not tied to the weather, so these farms can provide food all year round, and because they can operate on a limited footprint, they could be built right in the heart of the city. Urban farms can also grow plants using very little water (thanks to the lack of soil) and they produce considerably less waste during harvest than field-based farming. Smaller versions of these farms are being added to restaurants and schools, and flat-pack 'gardens' are being developed for apartment buildings. As part of a bigger solution, projects like these will have a big impact on the traditionally transport-heavy food chain.

Now before we move on to transport, I have to mention one question I've been asked repeatedly. **If we all became vegetarian, wouldn't that benefit everyone?** We already learned that vegetable production is less energy-intensive than meat production, but I suspected there was much more to this argument, so I asked Dr Kerr. He pointed out that farmers in developing nations are vital to the food production chain, and for them, a cow is more than just a walking hunk of meat: 'It is likely to also be a source of milk and fertiliser [in the form of dung], and the means by which a farmer brings their produce to the nearest town or city.' As well as that, it's worth reminding ourselves that in the UK, the grasslands our cattle graze on aren't fertile enough to grow crops on anyway – so we're not necessarily choosing meat over vegetables. As Dr Kerr said, 'The Scottish climate and our cattle are a great way to turn grass into food for humans.'* These grasslands have another role too: they are key habitats for the pollinators such as bees that our crops depend on. So taking all animals out of the food chain may not be a smart (or realistic) proposition. However, all of those I spoke to agreed that we absolutely need to cut down on our consumption of beef. So, there you go.

---

* The same argument is absolutely *not* true for those vast cattle complexes that feed soya and other products to cows. These make no ecological sense whatsoever.

Chain

Away from the farm – urban or otherwise – there are changes afoot in the way our food gets to us, and unsurprisingly, data is at the heart of it. Huge ports such as Rotterdam and Sydney have already begun to move to fully automated systems on-shore. There, automated straddles (small, box-shaped cranes on wheels) remove containers from trucks and drive them to a pre-specified storage bay, all without human intervention. They navigate using similar technologies to driverless cars: lasers and radar. Offshore activities are undergoing changes too. The cargo ships of today already rely heavily on computer control, but tomorrow, they may find their way to port entirely without a crew. The first potential benefit to this is safety. Just as on the road or in the air, most accidents at sea are caused by human error. In theory at least, by removing the humans, you drastically reduce the risk. The second benefit is fuel-related – not because crew members are heavy, but because, as naval architect Paul Stott told me, 'The fuel consumption of a ship is roughly proportional to its speed raised to the power of between three and four, depending on the hull shape.' In short, the faster a ship travels, the more fuel it uses. But slower ships mean longer journeys, which are significantly more difficult for the crews on board. In 2015, the marine division of Rolls-Royce announced a huge research project focused on designing slower crewless cargo ships, referred to as 'ghost ships' by pretty much every journalist. Working with a number of universities, ship designers and satellite companies, this project is looking into the technologies needed for navigation (a mixture of GPS, radar and video imaging), communications and remote control (mostly satellite-based).

These technologies are either already available, or in the final stages of development, but other challenges remain. Just as with the driverless cars we met in Chapter 5, when automated systems and humans occupy the same space, we may have problems. You see, we don't really follow the same rules. Everything is black or white, yes or no, stop or go for a computer. In contrast, humans exist in shades of grey – we can use our experience, instinct and judgement to resolve a situation. I'm not saying that one is better than the other, but to make autonomous systems work for us, we need to understand that difference. The computers aren't going to 'take over' – they can only do what we teach them to do – but realistically, this move will displace jobs. Being part of a culture that values trade unions, I totally understand that worry. All I can say is that technology has always had an impact on the jobs market, and the jobs themselves have changed in response. The term information and communications technology (ICT) basically didn't exist 50 years ago, but as of 2015, 7.7 million people were working in that sector in the EU alone. From everyone I've spoken to, it seems that this level of autonomy is not a pipe-dream. We may get to a stage (even within 20 years) when containers of exported goods will be driven to a city's port on driverless trains, loaded by robotic cranes onto massive, crewless cargo ships and sent on their way, all with little or no human interaction.

A logistics-based technology that makes my inner steampunk nerd very happy is the next generation of cargo blimps. Dr Tom Cherrett from the University of Southampton assured me that 'freight airships are being developed right now'. One of those leading the way is Kazakhstan-born engineer Igor Pasternak. His first prototype (launched in 2013) was 81m (266ft) long, with a rigid skeleton made from lightweight aluminium panels and carbon fibre trusses. The skeleton is covered in a shiny skin made from polymer materials to form the ship's gas bag (officially called an envelope). This might seem a bit crazy, especially when we cast our minds back to the airship disasters of the 1930s, but there are some benefits – airships can take off vertically (minimising the required launch site) and can carry huge amounts of cargo without burning fossil fuels. The issue with wrangling the airship to the ground may even be one step closer to being solved. Airships float because they are filled with gases that are lighter than air (historically, helium), making them very tricky to land safely.[*]

---

[*] There is a general shortage of helium gas, so hydrogen (produced with Chapter 2's electrolysers) has been suggested as a substitute.

Pasternak found a way to compress the helium in the airship's envelope by pumping air into it to make it heavier.* So how far away are they? Well, if production estimates are anything to go by, we may see them in the skies above our cities by 2025. Having always hoped for a Victoriana-type future, I must admit, I'm cheering for these airships.

On a smaller (and frankly, less epic) scale, sensors are also going to have a huge impact on the way we buy, ship and package our goods. Nanotechnology is already very popular in food and drink packaging, with its global market estimated to be worth $6.5 billion (£4.6 billion) in 2013. Traditional materials including paper and glass are still in use, but polymer films that contain nanocrystals of starch, silica or cellulose are becoming increasingly popular. They act as barrier materials for foods, keeping oxygen and humidity out and allowing food to stay fresher for longer, and they are biodegradable. Wider than that, it's now entirely possible to monitor and control the environment inside a shipment of food, from thousands of kilometres away. The technologies that do this include RFID, or radio-frequency identification, which uses a small computer chip surrounded by a coil of wires (a radio antenna) to uniquely mark an item. When combined with radio-based tracking, RFID tags can be used to pinpoint individual objects, and low-cost digital memory means they're getting smarter too. Conductive inks and printed batteries are also being developed that may soon lead to RFID tags that are printable, chipless and cheap.

Something that combines this data-rich approach to supply with the indoor farms we discussed earlier, is being spear-headed by Dr Caleb Harper from the MIT Media Lab. Described as 'an operating system for the farm of the future', CityFARM analyses absolutely everything about a plant's growing environment, from the pH levels in the water (how acidic it is) to the CO_2 content in the air around it. But here's

---

* Submarines use a similar approach to control their elevation – to sink, they allow water to fill their ballast tanks, and to float they fill them with air.

the cool thing. All of this data is then made publicly available, for free via a 'library of climate recipes'. From this, users can find the settings needed to grow the perfect tomato, pepper or head of broccoli in small, self-contained facilities anywhere in the world. Harper's main goal is to inspire a new generation of urban farmers, to encourage everyone to manage at least part of their own food production. According to *Wired*, Harper's systems are 'already being trialled across the world, from San Francisco and Detroit to Dubai, Chennai and Hong Kong'. Being a fan of both great technology and open source data, I will absolutely be keeping an eye on this project.

Connected systems like these only make sense if they form part of a much larger network. Thanks to the ever-decreasing cost of sensors, processors and internet access, that idea is fast becoming reality. It's time we talked about the Internet of Things.

Wireless

I will begin with a mini-rant. The term Internet of Things (IoT) is bandied around a lot, but I am not a fan. It sounds like it has a specific meaning, but it really, really doesn't. I very much view it as a cringe-inducing marketing term, and from my chats with people actually working in the sector it's not all that popular there either. However, because we're all being bombarded with it, and it neatly covers a lot of the tech I want to talk about, I'm going to use it anyway. But I just ask that you don't judge me for it, OK?

Anyway. IoT broadly refers to the concept of getting everything connected, both to the internet and to each other. If an object has an on/off switch, it could potentially be connected to the internet. So far, some of these ideas have seemed a bit silly – for example, the smart fridge that texts you when its cameras see you're out of milk – leading many to say that it's just information for information's sake. There is much more to it than that, as Dr Alex Bazin, Head of IoT at Fujitsu, told me: 'The idea at the heart of the hyperconnected era is that by using data and tech, you can have bespoke products at a mass-produced cost. A good example of this is the smartphone. No two are alike, even when they're identical models. Apps and tools allow a user to totally personalise it.' And you don't need me to tell you how much of an impact the smartphone has had on city living.

Within the next few years, IoT could utterly change the morning routine. The alarm that wakes you might also alert your shower and your coffee machine to turn on. Software that brings together weather information, traffic alerts, transport timetables and your calendar will be able to plan the best route to your office. For businesses, sensors that monitor buildings and vehicles could schedule their own maintenance. Crops could notify an urban farmer when they're ready to be harvested, and bins could tell the local council when they are full. In autonomous systems, we'll see constant flows of information using IoT – driverless cars will communicate with each other, the road and the traffic signals, all without you realising it. This might sound rather far-fetched, but *the technology is already here*. There are just a few gaps that still need to be bridged.

If you're wondering how far it could go, the answer is, nobody knows. There's very little consensus on the scale of IoT use in cities, even on a short timescale. IT analysts Gartner predicted that by 2020, 25 billion things would be connected to the internet, but Cisco upped that to 50 billion. Morgan Stanley tops the list of IoT optimists – they once predicted the number of internet-connected things to reach 75 billion by 2020, equivalent to 11 online devices for every person on Earth. Whatever the final tally, with everything transmitting data all the time, our data networks will take a hammering. Just like water and waste pipes, networks are designed for a certain capacity (called bandwidth). Once you surpass that, things get messy, so we're going to see some big changes. Instead of today's simple sensors, tomorrow's will be more like miniaturised computers – they'll be able to process data locally and only send the essential stuff to where it's needed. Even with that, some networks just won't be reliable enough for IoT, so other radio-based technologies, independent of Wifi, 3G and fixed-line internet, are beginning to make their mark.

As well as bandwidth, all of these systems will need something even more important: power, and it may not come from the wall-socket. In 2015, a team of computer scientists and electrical engineers from the University of Washington announced that they had used Wifi to deliver electrical power to a range of devices, and to charge batteries from over 8.5m (28ft) away.* Earlier we talked about how Wifi uses radio waves to transmit data. Using a small electrical circuit, it's possible to convert these radio waves into electrical energy. The researchers added this circuit to a temperature sensor and placed it near a standard Wifi router that transmits radio waves. The voltage they measured appeared in bursts – it was only there when the router was sending or receiving data. So they tricked the router into transmitting junk information on other channels whenever it was not being used. This produced a small, continuous voltage that was sufficient to power tiny cameras and battery chargers. The team also demonstrated their system, called Power over Wifi (PoWiFi) in six urban homes, and showed that harvesting power had no impact on the communication performance of the router. *MIT Technology Review* suggests that PoWiFi could be 'the enabling technology that finally brings the Internet of Things to life', and I have to say, I'm inclined to agree with them. They're not the only ones investigating this kind of technology – Nick Chrissos from Cisco told me about the fascinating work they're doing on delivering electrical power to streetlamps using internet cables, which he described as a 'step-change in the evolution of utilities'. Watch this space.

Of course, the other major issue for anything IoT-related is **security**. Data sharing and privacy are already in the headlines; once we get to a stage where tens of billions of objects are potentially hackable, it'll be a whole new ballgame. I don't mean to totally freak you out, but very few IoT

---

* Removing the need for wires in both power and communication was a long-held dream of Nikola Tesla.

products have in-built security – they're really only as secure as your home network, which isn't very secure at all. In recent years, a number of white-hat (*i.e.* friendly) hackers have shown that home thermostats, baby monitors and TVs are all easily tapped into. For now at least, there's very little valuable data being shared on these systems, but as soon as there is (say in driverless cars) we'll be in trouble. We're at a critical stage in the evolution of the Internet of Things – the potential applications are unbelievably exciting, but the security concerns are real and growing. The key is to stay one step ahead, and for that we'll need lots of programmers and computer engineers. Hint to careers counsellors / students.

Device

There is one thing that has enabled all of the technologies we've discussed in this glance into the future: the miniaturisation of computer components, as described by Moore's Law.* In 2015, IBM's research arm announced that they'd produced a computer chip with individual components (transistors) just 7 nanometres wide – that's about 10,000 times thinner than a human hair. But there's lots of research that's still in the lab that may make IBM's transistors look huge. One paper that caught my eye came from the Paul Drude Institute for Solid State Electronics. Working with researchers in Japan and the US Naval Research Laboratory, the team created a transistor from a single organic molecule surrounded by a small number of indium atoms. Traditional transistors can only be either on or off, but because of its size, a molecular transistor can tap into a quantum effect called entanglement – this transistor could be on, off, or both at the same time! Professor Andrea Morello from the University of New South Wales recently demonstrated that this entanglement exists in a silicon microchip. And why should we care? How about

---

* Moore's Law is less of a 'law' and more of an aspiration or observation. Intel co-founder Gordon Moore predicted that the number of transistors that could fit on a computer chip (a good measure of that chip's performance) would double every two years.

because it could utterly change the face of computing? As Professor Morello told me: 'Quantum computers can be exponentially more powerful than classical ones, because they make use of quantum entanglement to expand the vocabulary of computer code they can handle. It's as if, using the same 26 letters of the alphabet, we could suddenly use a vocabulary of billions of words instead of just a few hundred thousand.' For Morello, this has widespread implications: 'For certain problems, such as molecular simulations and perhaps drug design, using this extra quantum code can provide a formidable advantage.'

So while future computers may be rather different from what we have today, closer to home there's another issue with our love of gadgets. All of those we use depend on lanthanides (also known as rare-earth metals), which are commercially mined, mainly in China. The first thing to say is that these metals are neither rare nor 'earthy' – in reality, they exist in small quantities in many common rock types.* There are lots of rare-earths, and if you live in a city, I guarantee that you use at least one several times a day. Lanthanum is found in countless batteries in hybrid cars, europium makes the light from white LEDs look a little warmer, cerium oxide is used to polish smartphone screens and neodymium can be found in headphones, hard drives and wind turbines. Our modern world has been built on these metals, but it's on rather shaky foundations.

Extracting rare-earth metals from the ground is extremely energy-intensive, and the waste products can be highly toxic and polluting. Because we love to frequently upgrade our tech, tonnes of these useful metals lie in rubbish dumps. What can we do? One of the issues is that these rare-earths never appear alone; they must be extracted from a compound. A 2014 study by researchers at Leiden University suggested that **recycling** might be possible. They calculated that extracting neodymium from old hard drives would use 60 per cent less

---

* The name originally dates from the eighteenth century, and it just kind of stuck.

energy than mining the virgin metal, but offered no practical route to doing it. In August 2015, the Ames Laboratory announced that they'd developed a process to extract rare-earth metals from commercial electronic devices. Their system works by breaking the chemical bonds between atoms and forcing the rare-earth metals to bond to other, more easily removed, materials. They're currently working on commercialising this system, so although it's a belated start, this scientific approach to recycling is really positive.

Pay

Now, the last topic I want to talk about combines the two things we city-dwellers never have enough of: time and money. It's clear that the way we pay for goods and services has drastically changed in the past two decades. Cheques are being phased out in most developed cities, and while cash remains the most popular form of payment, other options are rapidly catching up. In 2014 in the UK, more transactions were made with cards, smartphones and online than with notes and coins. So, electronic payments seem to be the future and I can't possibly talk about that without mentioning cryptocurrency.

Yes, it might sound like some sort of virus, but virtual money may well feature heavily in the city of tomorrow. The idea is that it makes money international – any currency can be converted into a virtual currency, and then transferred online to other users without the need for a centralised service (for example, a bank). Bitcoin is probably the most famous cryptocurrency, but it certainly isn't the only one – people are beginning to trade in virtual currencies, despite their fluctuation in price and the lack of agreement on whether they are legal tender.* A lot of the initial hype around cryptocurrency has petered out, but the technology that sits behind it (called the

---

* At the time of writing, very few countries had published specific regulations on the status of virtual or cryptocurrencies – no-one seems to agree on what these currencies actually are.

block chain) may well be more future-proof. It is compatible with existing financial systems and acts like a secure balance sheet of all transactions. It isn't stored in one central location though – instead, every node in the network keeps a partial copy of the transactions. This not only means that it's very difficult to defraud a system (because the records are permanent), but it could also speed up transfers and payment. Banks are very interested in adopting it. But in 2015, researchers from Boston University showed that this reliance on the internet, and more specifically, internet time, could cause problems. Like other financial systems, the block chain uses timestamps to track transactions, but because it depends on a network, an inaccurate (or hacked) clock could potentially jam it up, hugely undermining its security. There's lots of work still to be done.

Elsewhere in the urban future, time too has a role, and it takes us back to GPS. Mobile phones, financial systems, power distribution, agriculture and transport all depend on freely available signals from satellites to give us accurate time. But the fact that we depend on it so heavily is frankly terrifying. Dana Goward, President of the Resilient Navigation and Timing Foundation told me: 'The issue is that GPS is a single point of failure. If we were to lose it for any extended period of time, we'd be thrown back a hundred years.' GPS signals are worryingly easy to disrupt, as demonstrated by engineers at the University of Texas in 2015. Led by Professor Todd Humphreys, they repeatedly sent false GPS signals to a drone, deliberately sending it off course. A few years prior to that, the group carried out the same experiment on a ship, and they've also shown that time-based sabotage in financial trading could cost billions. Given our reliance on GPS, our changing relationship with money and the growing interest in using drones for deliveries, GPS spoofing is a huge concern.

But given how often we've discussed historic urban technologies in *SATC*, it seems entirely appropriate that it's a reinvention of a World War II shipping system that could offer the security blanket we need. Based on a network of low-frequency radio stations on land and at sea, eLORAN is the equivalent of a ground-based GPS network. It is currently in limited use in the UK, Ireland and the US, but with researchers gradually improving its accuracy, we're likely to hear a lot more about it in the coming years. As well as offering a backup to GPS, eLORAN has another bonus – its high-power signals are difficult to spoof, and they 'penetrate underground, underwater and indoors', said Goward, 'making it a valuable complementary technology to GPS'.

If this chapter has taught us anything, it's that there is so much more to a city than its physical infrastructure or its skyline. Arguably it is the invisible connections – trade, communications and food – that have had the most profound impact on daily urban life. And with the many challenges facing us, it's not easy to predict what will happen to them in the future. The one thing I'm sure of (and that I hope I've convinced you of) is that science, engineering and technology will shape the way we build and interact with our cities, as they have done throughout history.

At the heart of all this is something we haven't discussed at all in *SATC*, and that is the role of people. So, for the next few pages, we're going to cast aside our sceptic's hats, and say farewell to our experts. We will instead visit an imaginary future city, and see it through the eyes of one of its residents. In a sense, this will be a bit more science fiction than science fact, but I hope you enjoy the journey.

CHAPTER EIGHT

City

Morning

As I doze in the small hours, the rest of the city is alive with activity. Since large trucks were banned from the city centre during the day, most deliveries now happen between midnight and dawn. But thanks to advances in electric and fuel-cell vehicles (not to mention drones and airships), they're almost completely silent, which keeps noise levels in our neighbourhoods low. The widespread use of autonomous systems in shipping and logistics means that humans now play a supervisory role in making all this happen. Their skills and expertise are used to ensure the robots keep doing their thing. For my elderly neighbour, the lack of heavy lifting has allowed him to stay in the job he loves.

Heading towards dawn, the roads begin to empty and the landscape of the urban jungle slowly changes. I've seen images of cities that once shone so brightly at night that they could be seen from space. It's a bit different now. Sure, there are some cities that still like to put on a lightshow, but these days everything's based on efficient light sources that are often powered by energy harvested during daylight. They're smart too – thanks to their internet connection and multiple on-board sensors, they feed data into local weather forecasts, and can act as charging points for gadgets. The lamps along my street are powered via their internet cables, which made installing them a whole lot easier too.

I wake up to the sound of the radio – funny how this technology has stuck around for such a long time. I look at the alert on my phone to see if there are delays on my usual route because of over-running roadworks (some things never change). Of the other suggested routes, I select the train and my travel tag is automatically topped up with the correct fare.

It also pre-loads discounts at a coffee shop close to my first meeting. It knows me so well.

I am not a morning person, so I'm glad I don't have to think too hard as I crawl out of bed. When I reach the bathroom door, the shower automatically starts running, and the electrochromic glass cubicle turns opaque. Since we've become more aware of the demands on our water supply, we've definitely changed our habits. The shower is time-limited and uses less than half the water that 'power showers' used to. The water I'm using right now is heated mostly by the combined heat and power (CHP) system in the basement. We get the rest of our heating from the solar thermal system on the roof. A lot of the city's publicly owned buildings tap into a district heating system that utilises waste heat from a nearby nuclear power plant. I'm proud to say that we're close to being free of fossil fuels, and that other cities are swiftly catching up.

Today I'm due to meet some new colleagues, so I'll need to head into the office. With my route already planned, I take a short walk through the park to the train station, and board the high-speed maglev train. I'm a bit unusual these days because I still work in the old central business district. Most businesses have now set up offices close to residential areas, which has really changed the old idea of a neighbourhood. One of my friends has recently moved into a new district that includes not only his office and home, but also a school for his children and lots of outside space. This approach has also meant that, thanks to the evolving nature of work, the city no longer functions on the old 9–5 routine. Office hours are staggered across the city, and with low-cost internet available everywhere, most people can work remotely for at least a couple of days each week. Fundamentally it means that today's urban dwellers don't move like iron filings to the magnetic heart of the city – the flow of people is much more dispersed. The old streets that I'm walking down now play a fairly minor role in the workings of the city, but tourists still love them.

Anyway, once I arrive at the office in the workings of the city, discount coffee in hand, I meet my new teammates. After a productive morning of discussions, we take the (cable-free)

magnetic elevator down to the ground floor and head out
for lunch.

Afternoon

Food consumed, I just have one quick thing to do before my
next meeting – I need to send my designs to the commercially
run fabrication lab in the basement. It's still very much the era
of the maker, and I'm pretty nifty on a 3D printer. I don't
have time for it this weekend though, so I add it to the
communal task list – I'll pick my component up when I'm
back in the office next week.

It's a beautiful day, so given that the next thing on my list
is a conference call, let's go to the roof garden. We have a lot
more greenery in our cities these days, which has given us
cleaner air than any city had in the early years of the century.
Fossil-fuel vehicles are pretty rare now – with the cost of oil
so much higher than the alternatives, they just don't make
sense. As you look around the garden, you may be wondering
where the solar panels are; they're hidden in plain sight! Every
pane of glass in the building is coated in a thin-film material
that harvests sunlight. Though practically invisible to our
eyes, this glazing provides a huge chunk of the office's
electricity needs.

Today's electricity grid is quite different from what you're
used to, more collaborative I guess. There is still a national
network, but these days, there are many more local systems
that produce and distribute their own electricity, like the
building I'm standing on. In terms of natural resources, we're
lucky too – our city is on the coast and we get regular sunshine,
so wind and solar power work well. Our apartment building
has solar panels on the roof, and our south-facing balcony is
clad in them (amongst the plants!). We also contribute to a
community fund to lease a series of offshore wind turbines. In
our district at least, the bulk of our electricity comes from these
two sources. We have huge improvements in energy storage to
thank for a lot of this – without that, it would be impossible to
balance supply and demand. Both at a district level and at a

national level, the network is constantly monitored; at all times, we can see exactly how much electricity we're using and where it comes from. I love having this level of data at my fingertips.

My afternoon meeting is a call with a colleague in Argentina and another in China. Thankfully, our simultaneous translation system makes communication between us almost seamless. Because the terminal scanned my face when I first sat down, it knows who I am, so as soon as the call ends, any meeting actions will pop up in my calendar. In the elevator my phone beeps to say that there's a peak in energy demand in my neighbourhood (there must be an event on!), so I take the opportunity to sell some of our solar electricity back to the local grid.

I have another meeting in a different part of town in a couple of hours, so let's explore more of the building until then. First, to the toilets. As in most cities, we've had to rethink the treatment of waste. Buried deep beneath the building is a bioreactor that breaks down solid waste from the loos and food waste from the kitchens, and transforms them into methane gas. This gas is then used as the fuel in a small heat plant (similar to the CHP in my apartment complex). The building is literally heated by poo! Any leftover solids are heavily treated until we're left with the perfect fertiliser for the roof garden. All of the wastewater from sinks undergoes treatment too – some of it is redirected for use in the fab-lab, but the cleanest stuff is reserved for us all to drink. The building isn't self-sufficient yet, but it's getting there!

From the outside, too, it looks pretty cool – while most of the structure consists of solar glazing, the bottom four floors are made with a concrete that encourages the growth of microbes naturally found in the air. The thing I love about that is, as seasons change, so too does our building's appearance – from forty shades of green in summer to the rich, jewelled tones of autumn. Now that we're outside, let's head over to the Loop. This is the quickest way to get across town, and it looks like someone else has requested the same route, so I'll have a buddy for the journey. We both hop into the pod and we're on our way. The Loop is a network of plastic tubes, through which small pods can travel at high speeds. Our city was among the first to install the system, and to be honest, it's still a rather expensive way to get around. I think of it as a treat – a much speedier version of the black cab from your era.

We're meeting in the café of a nearby park. Although the park looks much the same as always, there are some subtle differences from what you're used to: all of the benches and tables have been 3D-printed from recycled plastic, and hidden sensors constantly monitor the temperature and humidity in the communal vegetable gardens. Our weather patterns show a lot of variability (the result of a long history of burning fossil fuels). In order to help scientists understand the patterns, environmental data is collected everywhere and fed into their models. A good use of sensors, if you ask me! Anyway, two cups of coffee later and a project plan in place, I can head home.

Evening

I walk to the nearest hub – my phone tells me that I'll arrive in perfect time to catch the direct train. Moving through the station, credit is applied to my travel tag automatically, as my footsteps help to power some of the floor-level lights. I walk past a wall made from the cutting face of the machine that once dug these tunnels; engineering really is everywhere in the city. I emerge onto the platform as the train arrives (hydrogen-powered this time), so I get on.

Once home, I do a little more work before heading to the top floor to start gathering dinner. Our farm was one of the reasons we moved to this building. Using very little water (treated wastewater, that is) and electricity only from renewable sources, the residents have created a paradise from where we get all of our vegetables. I select a few potatoes and move to the salad section. We've been following a growing recipe from another farm, and our crop yield is amazing – we see little or no waste. I have ambitions to grow my own coffee beans too, but that's a little harder than no-soil tomatoes.

Before I eat, I need to go for a run — I've been slacking. I upload the data from my most recent workout onto my watch, set the targets, and check my heart-rate before I go. I decide to take the route by the river — it's a real showcase for many of the technologies that make my city unique. The LEDs that light the way brighten slightly as I approach them, and the glow-in-the-dark plants are a fun addition too. The walls of the river are definitely worth a closer look, especially those panels covered in the work of local street artists. Creatively, they have free rein — the only rule is that they have to use air-purifying paint. Everyone wins! Other sections of the embankment are covered in native plants; they thrive with only the smallest amount of nutrients, so they don't need big boxes of soil. Looking up, I see the latest skyscraper is considerably taller than when I saw it last — construction is much faster than it used to be. My watch tells me I've hit my targets, so I cross the river and head back towards home for dinner.

It's Saturday tomorrow, and we have a road trip planned, so I need to check the booking for the car. Private cars are very rare indeed, reserved for the super-rich (or super-foolish). We share our car with three other families in the building. It's electric, so I make sure that it's fully charged and that the correct amount of credit has been taken. We're planning to upgrade to a fuel-cell car soon — we've been redirecting some of the excess electricity from our turbines to an electrolyser nearby, so our hydrogen fuel tank is filling nicely.

Weekend

Morning! We've had breakfast, a shower, are caffeinated and we're about to hit the road. Almost all cars now can operate fully autonomously, but I still really enjoy driving. Our city is gradually phasing out road signage — there's no need for it when your car can navigate itself — but for now, traffic lights are still in use at junctions. The car constantly communicates with sensors in the road, and the traffic data collected helps us to choose the quickest route. I see only a few other drivers with their hands on the steering wheel — the latest cars don't even include one. Once we escape the city and hit the motorway, I let the autonomous system kick in. I can sit back and enjoy the book I've been reading... the weekend is finally here.

I know that all of these changes might seem Utopian, and ridiculously futuristic, but the truth is, you already have access to everything you need to build this city of tomorrow... what's stopping you?

Further Reading

The hardest thing about writing a book is deciding what to leave out. If I'd included everything I found interesting, *SATC* would have been at least twice as long! So, the following is my attempt to sate your hunger for details. This is not a comprehensive reference library – it's more of a mismatched list of papers (those behind paywalls are marked with a '£'), reports, books and links. But I hope it will help you start your own exploration of the urban landscape. If there's something you're particularly interested in but you need some guidance on where to find it, please contact me on Twitter (@ laurie_winkless) and I'll do my best to help.

Chapter 1

- CTBUH's skyscraper centre: www.skyscrapercenter.com/
- The best textbook on materials science: Michael F. Ashby, Hugh Shercliff and David Cebon 2012 (2nd edition). *Materials: Engineering, Science, Processing and Design.*
- A good overview on the history of steel: www.worldsteel. org/steelstory/
- Robert Courland 2011. *Concrete Planet: The Strange and Fascinating Story of the World's Most Common Man-Made Material.*
- On the vanadium self-cleaning window: J. Zheng *et al.* 2015. $TiO_2(R)/VO_2(M)/TiO_2(A)$ multilayer film as smart window: Combination of energy-saving, antifogging and self-cleaning functions. *Nano Energy* 11: 136–45 (£).
- Sandra Manso-Blanco's 2014 PhD thesis. *Bioreceptivity optimisation of concrete substratum to stimulate biological colonisation.* www.goo.gl/qEDnBX
- All of the World Health Organisation's air quality reports can be downloaded from www.who.int/phe/ publications/
- Read more about 'In Praise of Air' here: www. catalyticpoetry.org/

- This paper provides a good overview of photocatalytic surfaces: N.S. Allena *et al.* 2008. Photocatalytic titania based surfaces: Environmental benefits. *Polymer Degradation and Stability* 93 (9): 1632–46 (£).

Chapter 2

- David MacKay 2008. *Sustainable Energy – Without the Hot Air*. Free to download from David's website of the same name.
- If you're in the UK, you can find data on every city's energy consumption by searching for 'DECC statistical data set'.
- For New York's energy usage stats: B. Howard *et al.* 2012. Spatial distribution of urban building energy consumption by end use. *Energy and Buildings* 45: 141–51.
- Alternatively, explore the map yourself: sel-columbia.github.io/nycenergy/
- H.M. Paul 1884. Edison's Three-Wire System of Distribution. *Science* 4 (94): 477–8 (£).
- UNEP 2015. *District Energy in Cities*.
- Learn more about wind turbine blade design from this excellent book chapter: www.gurit.com/files/documents/3_blade_structure.pdf
- The National Renewable Energy Lab's PVWatts is an amazing resource (and a lot of fun). You'll find it here: pvwatts.nrel.gov/
- F.R. Martins, S.L. Abreu and E.B. Pereira 2012. Scenarios for solar thermal energy applications in Brazil. *Energy Policy* 48: 640–9 (£).
- Environment America Research & Policy Centre report: J. Burr and L. Hallock Spring 2015. *Shining Cities – Harnessing the Benefits of Solar Energy in America*.
- R.R. Hernandez *et al.* 2015. Efficient use of land to meet sustainable energy needs. *Nature Climate Change* 5: 353–8 (£).
- F. Creutzig *et al.* 2015. Global typology of urban energy use and potentials for an urbanization mitigation wedge. *PNAS* 112 (20): 6283–8.
- N. Debbage and J.M. Shepherd 2015. The urban heat island effect and city contiguity. *Computers, Environment and Urban Systems* 54: 181–94 (£).

- A. Gouldson *et al.* 2015. Accelerating Low-Carbon Development in the World's Cities. *The New Climate Economy*.
- MIT Energy Initiative 2015. *The Future of Solar Energy*.
- J. Moon *et al.* 2015. Black oxide nanoparticles as durable solar absorbing material for high-temperature concentrating solar power system. *Solar Energy Materials & Solar Cells* 134: 417–24 (£).
- To learn more about the vast array of large-scale storage options (including hydropower), read the IRENA *Electricity Storage Technology Brief*, 2015.
- First of two papers on batteries: M.C. Lin *et al.* 2015. An ultrafast rechargeable aluminium-ion battery. *Nature* 520: 324–8.
- Second, from Samsung: I.H. Son *et al.* 2015. Silicon carbide-free graphene growth on silicon for lithium-ion battery with high volumetric energy density. *Nature Communications* 6: 7393.
- Quote from Professor Sheila Widnall: www.technologyreview.com/s/537721/bladeless-wind-turbines-may-offer-more-form-than-function/
- Quote from Professor Sungho Jin: jacobsschool.ucsd.edu/news/news_releases/release.sfe?id=1589

Chapter 3

- Peter Gleick 2011. *Bottled and Sold: The Story Behind Our Obsession with Bottled Water*.
- World Bank, Urban Development Series 2012. *What a Waste: A Global Review of Solid Waste Management*.
- G. Grass *et al.* 2011. Metallic Copper as an Antimicrobial Surface. *Applied and Environmental Microbiology* 77 (5): 1541–7.
- The US Geological Survey Water Science School is a great general resource for the wet stuff – have a look at their website: water.usgs.gov/edu/sitemap.html
- DEFRA and the Food and Environment Research Agency 2010. *The role and business case for existing and emerging fibres in sustainable clothing*.
- To find out more about the water footprint of different activities, start with the UN water and food security page: www.un.org/waterforlifedecade/food_security.shtml. You

could also look for any paper (mostly freely available) by Professor A.Y. Hoekstra. A good starting point would be the report he wrote for UNESCO: *The green, blue and grey water footprint of farm animals and animal products*, Vols 1 and 2. Another good option is the report from the Institution of Mechanical Engineers 2013: *Global Food: Waste Not, Want Not.*

- For wastewater and other globally relevant water research, look at UN-Habitat (unhabitat.org/) and the UN Environment Programme (www.unep.org/) websites – all reports are free to access.

- J.T. Powell *et al.* 2016. Estimates of solid waste disposal rates and reduction targets for landfill gas emissions. *Nature Climate Change* 6: 162–5 (£).

- K.C. Park *et al.* 2013. Optimal Design of Permeable Fiber Network Structures for Fog Harvesting. *Langmuir* 29 (43): 13269–77 (£).

- S.C. O'Hern *et al.* 2014. Selective Ionic Transport through Tunable Subnanometer Pores in Single-Layer Graphene Membranes. *Nano Letters* 14 (3): 1234–41 (£).

- W. Lei *et al.* 2013. Porous boron nitride nanosheets for effective water cleaning. Nature Communications 4 Article no. 1777.

- M. Bhattacharjee *et al.* 2015. Low algal diversity systems are a promising method for biodiesel production in wastewater fed open reactors. *ALGAE* 30 (1): 67–79.

- Y. Yang *et al.* 2015. Biodegradation and Mineralization of Polystyrene by Plastic-Eating Mealworms. Part 2. Role of Gut Microorganisms. *Environmental Science & Technology* 49 (20): 12087–93.

- Quote from Dr Craig Cogger: modernfarmer. com/2014/07/stink-human-poop-fertilizer/

- Quote from Dr Wei-Min Wu: news.stanford. edu/2015/09/29/worms-digest-plastics-092915/

Chapter 4

- Tom Vanderbilt 2009. *Traffic: Why we drive the way we do (and what it says about us).*

- The World Factbook from the CIA is available for free online: www.cia.gov/library/publications/the-world-factbook
- The Tacoma Narrows video is widely available on all video-sharing websites – it's a great example of how resonance can be destructive.
- To watch the video of phantom traffic jams building up in a closed racetrack, look here: www.goo.gl/lc4uoG
- C.H. Papadimitriou and J.N. Tsitsiklis 1999. The complexity of optimal queuing network control. *Mathematics of Operations Research* 24: 293–305 (£).
- M. Audo, *et al*. 2015. Subcritical Hydrothermal Liquefaction of Microalgae Residues as a Green Route to Alternative Road Binders. *Sustainable Chemistry & Engineering* 3 (4): 583–90 (£).
- There are lots of people working on variants of self-healing concrete – have a look for Hendrik Jonkers from Delft University, Bob Lark from Cardiff University and Professor Chan-Moon Chung from Yonsei University.
- B.J. Blaiszik *et al*. 2010. Self-Healing Polymers and Composites. *Annual Review of Materials Research* 40: 179–211.
- For more on composite materials for bridge decks, see the Institute of Bridge Engineering at the University of Buffalo.
- Halo was recently bought by Qualcomm and more details on the technology are widely available online.
- You'll find more information on standardisation of e-vehicle charging at the Intelligent Transportation Systems office: www.its.dot.gov/
- For solar-based roads and paths, again there's lots online – start with companies Wattway, SolaRoad and Solar Roadways.
- M. Jackett and W. Frith 2013. Quantifying the impact of road lighting on road safety – A New Zealand Study. *IATSS Research* 36 (2): 139–45.
- MIT AgeLab's suit is called AGNES – you'll find a great video about it on the MIT Video page: www.video.mit.edu/

- S. Box 2014. Supervised learning from human performance at the computationally hard problem of optimal traffic signal control on a network of junctions. *Royal Society Open Science* 1: 140211.
- Presentation on the Millau Viaduct bridge available on the Engineers Website: 'The Formwork to The Millau Viaduct', November 2007.
- Quote from Jairam Ramesh: timesofindia.indiatimes.com/india/If-there-is-a-Nobel-prize-for-filth-India-will-win-it-Jairam-Ramesh/articleshow/5251864.cms
- Quote from Nicola Davison: www.theguardian.com/cities/2015/feb/26/3d-printed-cities-future-housing-architecture

Chapter 5

- The US Environmental Protection Agency have collected a lot of data on fuel economy of various vehicles – you can access it all on their website: www3.epa.gov/otaq/
- The inventor of diesel, Rudolf Diesel, was an interesting character, right down to his mysterious death. If you can get your hands on the book by W. Robert Nitske and Charles Morrow Wilson, *Rudolf Diesel: Pioneer of the Age of Power*, it's worth a read.
- The Toyota, Tesla and BMW websites all include a lot of technical information on their hybrid and electric cars.
- Some information on Brazil's relationship with bioethanol is available on the International Energy Agency website (£): www.iea.org. However, the book *Biodiesel – Feedstocks, Production and Applications* is free to download from the INTECH website, www.intechopen.com, and is a good primer on the topic.
- You can find lots of information on the European Commission focus on renewables (and biofuels) on their website.
- L. de Schutter and S. Giljum 2014. *A calculation of the EU bioenergy land footprint*.
- The HyFive project has its own very shiny webpage: www.hyfive.eu/

- Lightweighting materials are all the rage for cars – the laser treatment project was led by the Fraunhofer Institute, but the work on aluminium is being carried out by individual car manufacturers. Mazda's bioplastic was announced in 2014.

- A. Elmarakbi 2015. A Short Overview of Graphene Nanocomposites for Automotive Structural Applications. *International Journal on Automotive Composites, Autumn Highlights*.

- M.A. Rahman *et al.* 2011. Development of a catalytic hollow fibre membrane micro-reactor for high purity H_2 production. *Journal of Membrane Science* 368: 116–23 ($\pounds$).

- There are so many battery-related papers in the literature, but I've listed just two here: Y. Liu *et al.* 2013. Feasibility of Lithium Storage on Graphene and Its Derivatives. *Journal of Physical Chemistry Letters* 4 (10): 1737–42 ($\pounds$); and Z. Favors *et al.* 2015. Towards Scalable Binderless Electrodes: Carbon Coated Silicon Nanofiber Paper via Mg Reduction of Electrospun SiO_2 Nanofibers. *Scientific Reports* 5: 8246.

- Read a free article on the energy-harvesting toy jeep by searching for *Rolling, rolling, rolling: harvesting friction from car tyres*.

- You can watch a video on the IBM/Eindhoven pilot by looking at the IBMBeNeLux YouTube channel.

- The Transportation Research Board has a huge amount of information on its website, some free, some requiring registration to access.

- The video of Andy Greenberg's SUV-hacking experiment is widely available on sites including Wired, Forbes and YouTube.

- Quote from Professor Ahmed Elmarakbi: www.sunderland.ac.uk/newsevents/news/news/index.php?nid=2810

Chapter 6

- There are countless books on London's Tube, but I'm a big fan of Paul Moss's book from 2014, *London Underground Haynes Manual*.

- The Crossrail project is expected to be completed in 2018. There's a lot of information on the project available on their website: www.crossrail.co.uk/construction/

- At the moment, Elon Musk's Hyperloop project is in its infancy, with not a lot of details available – but keep an eye on the SpaceX webpage.

- Learn lots more about the various 'Grades of Automation' available for metro trains on the International Association of Public Transport webpage.

- For more on the European Train Control System, head to www.uic.org/ETCS

- D. Fournier PhD thesis 2015. *Metro Regenerative Braking Energy Optimization through Rescheduling.* Paris Diderot University HAL Id: tel-01102408.

- S. Lu PhD thesis 2011. *Optimising Power Management Strategies for Railway Traction Systems.* University of Birmingham ID Code: 3091.

- N.J. McCormick *et al.* 2014. Assessing the condition of railway assets using DIFCAM: results from tunnel examinations. *6th IET Conference on Railway Condition Monitoring*: 1–6 (£).

- T. Moreno *et al.* 2014. Subway platform air quality: Assessing the influences of tunnel ventilation, train piston effect and station design. *Atmospheric Environment* 92: 461–8.

- Quote from the *Washington Post*: www.washingtonpost.com/wp-dyn/content/article/2006/08/04/AR2006080401755.html

- Quote from BBC: www.bbc.co.uk/news/magazine-32446717

Chapter 7

- The Marine Traffic website is amazing. You have been warned: www.marinetraffic.com/

- J. Tournadre 2007. Signature of Lighthouses, Ships, and Small Islands in Altimeter Waveforms. *Journal of Atmospheric and Oceanic Technology* 24: 1143–9.

- J. Tournadre 2014. Anthropogenic pressure on the open ocean: The growth of ship traffic revealed by altimeter data analysis. *Geophysical Research Letters* 41 (22): 7924–32.

- Marc Levinson 2008. *The Box: How the Shipping Container Made the World Smaller and the World Economy Bigger.*

- J.P. Rodrigue 2014. *Reefers in North American Cold Chain Logistics.* Hofstra University.

- Tom Jackson 2015. *Chilled: How Refrigeration Changed the World and Might Do So Again.*

- UN Food and Agriculture Organization (FAO) 2013. *Tackling climate change through livestock.*

- *Agriculture, Forestry and Other Land Use Emissions by Sources and Removals by Sinks (1990–2011).* FAO Statistics Division: ESS/14-02.

- D. Satterthwaite *et al.* 2010. Urbanization and its implications for food and farming. *Philosophical Transactions of the Royal Society B* 365: 2809–20.

- For a great primer on time, visit www.npl.co.uk/educate-explore/what-is-time/

- The UCS Satellite Database is available to download from their website: www.ucsusa.org/

- Michio Kaku 2012 (2nd edition). *Physics of the Future.*

- J. Cook *et al.* 2013. Quantifying the consensus on anthropogenic global warming in the scientific literature. *Environmental Research Letters* 8: 024024 (7pp).

- All of XKCD is wonderful, but it is this one that feels most relevant here: www.xkcd.com/1601/

- M.L. Jat *et al.* 2011. Layering Precision Land Leveling and Furrow Irrigated Raised Bed Planting. *American Journal of Plant Sciences* 2: 578–88.

- Watch a video of Caleb Harper talking about CityFARM on the Wired UK YouTube channel.

- V. Talla *et al.* 2015. Powering the Next Billion Devices with Wi-Fi. arXiv: 1505.06815.

- Listen to an NPR radio piece on the IBM 7nm transistor on www.npr.org

- J. Martínez-Blanco *et al.* 2015. Gating a single-molecule transistor with individual atoms. *Nature Physics* 11: 640–4.

- BBC Future published a series of brilliant articles on rare-earth extraction – start with *The dystopian lake filled by the world's tech lust* at www.bbc.co.uk/future

- B. Sprecher *et al.* 2014. Life Cycle Inventory of the Production of Rare Earths. *Environmental Science & Technology* 48 (7): 3951–8.
- A. Malhotra *et al.* 2015. Attacking the Network Time Protocol. Available by searching on the Boston University Computer Science webpage: www.goo.gl/TiXe4l
- T.E. Humphreys 2011. *GPS Spoofing and the Financial Sector, a short introduction*, available at www.goo.gl/zMZqcg
- M.L. Psiaki and T.E. Humphreys 2016. GNSS Spoofing and Detection. Talk delivered at the International Symposium on Navigation and Timing, 2015. Published in *Proceedings of the IEEE*.
- To learn more about eLORAN, a good starting point would be the Resilient Navigation and Timing Foundation webpage: www.rntfnd.org
- Quote from *Wired*: www.wired.co.uk/news/archive/2015-10/15/caleb-harper-urban-farms-mit-wired2015
- Quote from *MIT Tech Review*: www.technologyreview.com/s/538031/first-demonstration-of-a-surveillance-camera-powered-by-ordinary-wi-fi-broadcasts/

General

These are not science books, but if you are a fan of visuals (infographics or colouring-in), these city-related titles might be of interest:

- James Cheshire and Oliver Uberti 2014. *London: The Information Capital.*
- Frank Jacobus 2015. *Archi-Graphic: An Infographic Look at Architecture.*
- Paul Knox 2014. *Atlas of Cities.*
- Ton Dassen and Maarten A. Hajer 2014. *Smart About Cities: Visualising the Challenge for 21st Century Urbanism.*
- Steve McDonald 2015. *Fantastic Cities.*
- Mister Mourao 2016. *Fantastic Cityscapes.*

Acknowledgements

Writing this book has made me realise just how wonderful people are. To everyone who gave up their time to help me in any way, thank you!

Those I interviewed or quoted: Bill Baker, Ron Slade, Laurie Chetwood, Dan Safarik, Sandra Manso, Phil Brown, Jason Garber, Tony Ryan, Daniel Schwaag, Ruth Finkelstein, Con Doolan, Senan McGrath, Win Rampen, Kirsten Dyer, Jamie Taylor, Chris Case, Sheila Widnall, Paul Law, Rory Mortimore, Stan Golunski, Sian Thomas, Sandy Lawson, Mike Jackett, Simon Box, Glynn Barton, Rajagopalan Vasudevan, Brian Maroney, Alan Turnbull, Geoff Holmes, Lindsay Chapman, Tom Vanderbilt, Benjamin Seibold, Richard Buswell, Nick Chrissos, Gareth Hinds, Ben Kingsbury, Roger McKinlay, Finn Coyle, Gary Filbey, Ahmed Elmarakbi, Rolf Dollevoet, Linda Miller, Steve Boyle, Alan Baxter, Ailie MacAdam, Nick McCormick, Siv Bhamra, Mick Lowe, Tom Cherrett, Alex Bazin, Paul Stott, Louise Manning, John Kerr, Patrick Gill, Rich Hunt, Andrea Morello, Dana Goward.

The many others who advised me, talked about their work, introduced me to others, sent me reports and papers, lined up interviews or fact-checked my ramblings: Chris Dulake, David MacKay, Lily Riahi, Askwar Hilonga, Mike Gower, Neil Wood, Korak VanTuyl, Andy Knox, Mark Stewart, Jon Elwood, Donal Finegan, Philip Whittington, Fathi Tarada, Andy Alder, Neil Murray, Eoin Ready, Ander Maderiaga, Ben Piper, Arthur Turrell, Simon Austin, John Preston, John Rowland, Colin Porter, Chris Speed, Paul Coxon, Peter Whibberley, Stephanie Fontaine, Adrienne Jacobs, Charlotte Schofield, Lisa Berardi Marflak, Andrew Hancock, Kimberley Steed-German, Sophie Millward, Emma Silke, Jane Griffin, Tom Canning, Leah Robinson-Leach, Tamara Salhab, Dagmar Dua, Tom Lawson,

Leanne Bell, Tom Warren, Emma Sinclair, Sandeep Dhillon, Proof Communication, Peter McLennan, Andrew Dempsey.

My wonderful chapter reviewers: Nic Harrigan, Lisa Martin, Leanne Archer, Lisa Dunbar, Ciaran Dunbar, Stewart Barker, David Lamb, Kimberley Steed-German, Nick Kennedy, Lindsay Chapman, Gareth McGrath, John Englishby, David Sutton, Nic Galtié Slavin, Dirk Schoellner, Keith Lawrence, Maria Fernandes, Alex Cuenat, Peter Woolliams, Gillian Lewis, Stevyn Colgan, John Molloy, Brian Durcan, Lauren Petrie, Richard Jackett, Jessica Marshall, Andrew Hanson.

And special thanks must go to the team at Bloomsbury: Sigma-Daddy Jim Martin, who continually displayed a frankly worrying level of trust in my ability to do this, and the actual boss, Anna MacDiarmid, who held my hand while making the whole thing come together.

Index